T0334546

Industrial Construction Estimating Manual

Industrial Construction Estimating Manual

Kenneth Storm

Gulf Professional Publishing
An imprint of Elsevier

Gulf Professional Publishing is an imprint of Elsevier
50 Hampshire Street, 5th Floor, Cambridge, MA 02139, United States
The Boulevard, Langford Lane, Kidlington, Oxford, OX5 1GB, United Kingdom

Notices
Knowledge and best practice in this field are constantly changing. As new research and experience broaden our understanding, changes in research methods, professional practices, or medical treatment may become necessary.

Practitioners and researchers must always rely on their own experience and knowledge in evaluating and using any information, methods, compounds, or experiments described herein. In using such information or methods they should be mindful of their own safety and the safety of others, including parties for whom they have a professional responsibility.

To the fullest extent of the law, neither the Publisher nor the authors, contributors, or editors, assume any liability for any injury and/or damage to persons or property as a matter of products liability, negligence or otherwise, or from any use or operation of any methods, products, instructions, or ideas contained in the material herein.

British Library Cataloguing-in-Publication Data
A catalogue record for this book is available from the British Library

Library of Congress Cataloging-in-Publication Data
A catalog record for this book is available from the Library of Congress

ISBN: 978-0-12-823362-7

For Information on all Gulf Professional Publishing publications visit our website at https://www.elsevier.com/books-and-journals

Publisher: Joe Hayton
Acquisitions Editor: Katie Hammon
Editorial Project Manager: Michelle W. Fisher
Production Project Manager: Sojan P. Pazhayattil
Cover Designer: Matthew Limbert

Typeset by MPS Limited, Chennai, India

Working together
to grow libraries in
developing countries

www.elsevier.com • www.bookaid.org

Contents

Preface

This first edition of *Industrial Construction Estimating Manual* provides a detailed estimating method using the unit-quantity model to prepare construction estimates for process plants using computerized estimation. Most estimating methods are qualitative and are not based on historical data. The unit method uses historical and quantitative data that leads to a cost driver easily understood. The unit method is used extensively in construction estimating. This method is extended by developing the unit method and historical man-hours into the unit-quantity model.

When given historical data from previous similar work task, the estimator uses the unit-quantity model to calculate cost and man-hours by category and produce reliable detailed cost and man-hour estimates. This detailed estimating method is the most accurate and timely method and when the computer does the manual and repetitive work, the estimator can spend more time on quantity takeoff and use graphical, statistical, or mathematical methods. This will enable the estimator to set up company cost and man-hour databases on a computer-estimating system that are timely, accurate, reliable and enable the estimator to verify historical data, using statistical methods and to compile bid proposals, RFP's and change orders. Companies must develop and maintain their own historical man-hour databases for earthwork, foundations, structural and miscellaneous steel, process equipment, piping, instrumentation, boilers, tanks, and pipelines and they need to be revised continuously.

The purpose of this manual is to enable the reader to use detailed estimating using the unit-quantity model to estimate industrial process plant construction work. The first chapter, in the manual, is an introduction to construction estimating. The chapter provides the reader basic information on construction databases, job cost by cost code and type, productivity measurement, and development of detailed estimating using the unit-quantity model. Successful bidding ensures the cost and material requirements associated with the installation of the materials. Chapter 2, Construction Material, deals with construction materials and provides sample material takeoffs. Estimating data systems use historical data from quantity takeoff quantities and man-hours for estimating new projects of similar equipment or systems. Chapter 3, Construction Database System, will enable the reader to develop labor tables that can be used to set up and implement a man-hour database system using comparable cost and man-hour data to estimate future projects. Construction labor estimates, covered in Chapter 4, Construction Labor Estimate, provide the elements of the construction-work estimate and illustrate estimate worksheets using the unit-quantity model. In Chapter 5, Computer Aided Estimation, we consider computerized estimating as a standard tool in the construction industry to automate and control estimating spreadsheets used to estimate more effectively and efficiently. Chapters 6−9 provide the reader labor estimates based on detailed estimating using the unit-quantity model to estimate construction work in industrial process plants.

The four chapters, six through nine, provide labor estimates for the following process plants:

- Chapter 6, Combined Cycle Power Plant (1 × 1) Labor Estimate
- Chapter 7, Gasifier Labor Estimate
- Chapter 8, Refinery Equipment and Storage Tank Labor Estimates
- Chapter 9, Circulating Fluidized Bed Boiler Labor Estimates

Then Chapter 10, Bid Assurance, introduces the reader to the analysis of estimates. The purpose of Bid Assurance is to provide the reader methods, and techniques that optimize the bid and regulate cost to match the estimate. An estimates accuracy, reliability, and consistency are important. If the bid is successful then there is an opportunity to verify the estimate using statistical analysis. Chapter 10, Bid Assurance, describes the unbalanced bidding strategy, estimate errors, estimate analysis, and assurance.

Chapter 11, Detailed Estimating Applications to Construction, provides practical applications to construction using Excel statistical and mathematical functions that model the work of construction.

The manual has been written to appeal to engineering, technology, construction estimating, and management settings. An effort has been made to provide the reader methods, model's procedures, formats, and technical data for preparing industrial process plant estimates using the unit-quantity model for detailed estimating. This manual will be an excellent reference for readers engaged in the construction industry.

Chapter 1

Introduction to construction estimating

1.1 Introduction

This chapter provides the reader methods, models, procedures, formats, and technical data for developing industrial process plant construction estimates. Detailed construction estimates are critical for engineering firms, contractors, and subcontractors to prepare accurate, reliable, verifiable, and consistent construction estimates. The term industrial process plants include firms involved with the construction of power plants, petroleum plants, petrochemical plants, and manufacturing plants. These process plants have a common reliance on process flow diagrams (PFDs), piping and instrument diagrams (P&IDs), vendor equipment scopes of work, and erection sequences as primary scope-defining documents. These documents are key deliverables in determining the scope of work and erection sequence and provide the level of detail required for a detailed work estimate. The detailed estimate is developed by combining the unit method and historical man-hours with the unit-quantity model to estimate industrial process plant construction work. Estimates for industrial process plants include mechanical process equipment and involve piping, instrumentation, structural and miscellaneous steel, and civil work. The methods, procedures, formats, and technical data throughout this manual provide information for compiling detailed construction estimates. The data must be modified as the user's experience and operational situation suggests. Man-hours are based upon direct labor and do not include indirect and overhead labor. The man-hours in this manual are based on each process plant's scope of work and erection sequence. Standard man-hours have been verified by regression models and adjusted for idle time, fatigue, and delays (PF&D). The man-hours are competitive in all geographical areas in the United States and should be factored for each contractor workforce productivity, weather, and any factors that may affect field craft productivity.

1.2 Types of construction estimates

Levels of accuracy for construction estimates vary depending on the stages of project development. Accuracy of construction estimates range where no

Industrial Construction Estimating Manual. DOI: https://doi.org/10.1016/B978-0-12-823362-7.00001-6

information is available and the cost estimate is expected to be less accurate to when the scope of work is clearly defined and the erection sequence identified. Many types of cost estimates are used at different stages of the project development, and the cost estimate will reflect the information available at the time of estimation. Cost estimates can be best classified into three levels of estimates according to their functions. The focus of this manual is to provide estimators and engineers a detailed estimating method using the unit-quantity model to prepare construction estimates for process plants using computer-aided cost estimation.

1.2.1 Levels of estimates

1. *Preliminary estimates (or conceptual estimates)*
  a. Preliminary plans and specifications.
  b. Little or no detail.
  c. General description of project.
2. *Detailed estimates (or definitive estimates)*
  a. A detailed work estimate is the most accurate and timely work estimate.
  b. Duplication of design enables an estimator to define and set up work scope and erection sequences for field construction work.
  c. The estimate requires historical data that has been collected, organized, and verified by statistical analysis.
  d. The data must be updated with respect to changes that will incur.
  e. The detailed estimate enables the contractor to schedule the work and complete construction successfully.
  f. Trends in man-hour units and cost will occur over time and forecasting future cost will be required.
  g. Time series—use moving averages and exponential smoothing to forecast changes over time.
3. *Engineer's estimates*
  a. Based on plans and specifications.

1.3 Models

Statistical and mathematical models are used in construction estimating to enhance construction cost analysis by graphic and analytical techniques. Statistical and mathematical models are easy to manipulate and can be used:

- to estimate the unit-quantity model,
- to explain the real situation,
- to provide a quick and inexpensive estimate,
- to use computers to handle regression models,
- to allow discovery of pertinent variables and comparison, and
- to use computer Excel spreadsheets and Excel functions to solve linear regression equations and statistical measures of reliability.

1.4 Labor productivity analysis

1.4.1 Labor

Labor is the most important item in estimating construction work. Craft labor is classified into direct−indirect and union−nonunion. Labor productivity is concerned with direct craft labor. Craft labor time means the craft is working in the field on construction activities. Indirect craft labor is supportive of direct craft labor.

Labor cost is defined by the following formulas:

- Man-hour = time × quantity (refer to the unit-quantity method)
- Labor cost = man-hour × labor rate (refer to unit method)

where time is in hours per unit and the labor rate is expressed in dollars per hour.

Once time values are known for a construction task, they are multiplied by the quantity. Time may be individuals or for crew work, and it is based on the construction task. Time is expressed relative to a unit of measure, such as LF, EA, SF, and ton. The unit of time may be a minute, hour, day, month, or year.

1.4.2 Labor man-hour

- The "man-hour" is dependent on the historical value of time spent doing construction activities.
- This basic unit is defined as one worker working for 1 hour.

Examples of man-hour units:

- Welding butt weld, carbon steel, arc-uphill, $0.562'' \geq WT \leq 0.688$, 1.05 MH/diameter in.
- Erect structural steel; >100 ton, X heavy—80−120 lb/ft., 11.8 MH/ton
- Structure backfill and compact—loader and wacker, 0.60 MH/CY
- Fabricate, install, and strip foundation forms—1 use, 0.30 MH/SF

1.4.3 Cost analysis

- Man-year is 52 weeks at 40 h/week, equal to 2080 hours.
- Man-month is 173.3 h/month $(40 \times 52)/12 = 173.3$.

Man-hours are used for estimating industrial construction work. Man-hours are effective when measurements and analysis of worker's time with respect to constructive and nonproductivity effort and idleness have been made. Man-hours have interface dependencies and must be based on quantitative measurement supported by historical data that has been verified by graphic and statistical analysis.

1.5 Data collection and regression analysis

Historical data is cost coded and collected in field construction. The data is collected from similar projects and is used as detailed backup for the estimate. When using cost data the estimator must be aware of the source of the data and make adjustments when necessary. If historical data is used, the data may not be accurate, and not applicable due to escalation. It must be reviewed and verified using regression analysis and, if necessary, use time series to account for escalation before the data is used in the estimate.

1.5.1 Construction database

Using software and technology the contractor can build and maintain an accurate and reliable craft labor database as well as create simple and easy to use Excel spreadsheets that anyone in the company can use to access, edit, and manipulate the craft labor. The estimator collects historical estimating data from previous projects to develop the estimating database for process equipment, piping, civil, and structural steel installed from previous projects.

1.6 Quantity takeoff

The quantity takeoff is developed during the bid preparation and quantifies the materials required to complete the project. Steps in creating the material estimate depend on the accuracy of the material estimate derived from the material takeoff (MTO), also known as the quantity takeoff. The takeoff refers to taking each of the required materials off from the drawings for the project. The takeoff is a count of how much material will be required for the project to complete it per the specifications provided in the bid package.

1.6.1 Material takeoff RP piping and supports

RP piping and supports					SB		SB		
Line no.	Material	Size	Sch/Thk	Pipe	SW	Valve	PS	Instrument
RP-50 RH desuperheater spray water	SA-106-B	1.5	80	65	25	5	3	
RP-51 reheater 1 to 2 drain	SA-106-B	2	80	40	12	2	2	
1.5″ RP-50	SA-106-B	1.5	80	60	19	7	3	

(*Continued*)

(Continued)

RP piping and supports				SB		SB		
Line no.	Material	Size	Sch/ Thk	Pipe	SW	Valve	PS	Instrument
CRH boot drain	SA-106-B	2	80	20	4		1	
HRH boot drain	SA-106-B	2	80	20	4		1	
Preheater drains	SA-106-B	2	80	180	54	9	9	

1.7   Scope of work and erection sequence

Process plants identified by the process of the plant and the equipment, regardless of the manufacturer, have designs similar to the previous designed systems. The duplication of design enables the estimator to define the scope of work and erection sequence. The process plants have a common reliance on PFDs, P&IDs, vendor equipment scopes of work, and erection sequences as primary scope-defining documents. These documents are key deliverables in determining the scope of work and erection sequence.

1.7.1   Scope of field work required for HRSG SB code piping

Scope of work-field erection RP piping and supports
  RP piping and supports
  RP-50 RH desuperheater spray water
  RP-51 reheater 1 to 2 drain
  1.5″ RP-50
  CRH boot drain
  HRH boot drain
  Preheater drains

1.8   Coding

1.8.1   Job cost by cost code and type

Cost codes

1. Identifies a specific task within a job.
2. Cost type identifies specific cost within the activity, such as labor and material.

3. Cost code is used to describe the task and cost type to describe the man-hours associated with the task.

Coding advantages:

- Estimate preparation
- Data management

The company must develop a detailed code that has a structured scope of work and erection sequence corresponding to construction process of the company's construction activity.

The process plants are identified by the process of the plant, and the process provides the scope of work and erection sequence required to install the scope of work in a process plant.

The estimator develops the scope of work in the erection sequence to install civil, structural steel, and mechanical work in the field and set up cost codes to identify the construction task. Historical data is collected from field construction, and similarities are compared to actual cost and time spent on work activities with those of the estimate. Historical data collected for civil, mechanical, and boiler construction based on cost code/type for each task is averaged and summarized into a spreadsheet, and the productivity rate is determined. The rate is used for future man-hour analysis and estimating jobs that are similar. Productivity rates are entered into a craft labor database (man-hour tables), and the craft labor hours have been verified by graphic and regression analysis.

Industrial contractors develop their own specialized cost codes for material and labor cost expressed in unit of man-hours.

1.8.2 Table—jJob cost by cost code and type—HRSG RP piping and supports

Cost code	Type	MH
XXXXXX	RP-50 RH desuperheater spray water	116.2
XXXXXX	RP-51 reheater 1 to 2 drain	66.8
XXXXXX	1.5″ RP-50	105.4
XXXXXX	CRH boot drain	35.5
XXXXXX	HRH boot drain	35.5
XXXXXX	Preheater drains	244.7

1.9 Productivity measurement

Historical records provide the direct craft man-hour data for field installation of civil and mechanical work. Two methods for the measurement of construction time are used to code, collect, analyze, and compile the actual man-hour data in this manual.

1.9.1   Nonrepetitive one-cycle time study

Normal time is found by multiplying a selected time for the task or cycle by the rating factors.

$$T_n = T_0 \times RF$$

where T_n is the normal time (hours), T_0 is the observed time (hours), and RF is the rating factor and arbitrarily set (number).

Normal time does not include factors that affect labor productivity. Allowances for these factors are divided into three components: personal, fatigue, and delay (PF&D) process of timing the cycle:

- Idle time is excluded; craft takes breaks for coffee and rest room; allowance for personal is 5%.
- Fatigue is physiological reduction in ability to do work; allowance for fatigue is 5%.
- Delays beyond the worker's ability to prevent; allowance for delays is 5%.

Productivity time in the work day is inversely proportional to the amount of PF&D allowance, and the allowance is expressed as a percent of the total work day.

PF&D allowance is generally in the range of 10%−20%; allowance multiplier:

$F_a = 100\%/(100\% - PF\&D\%)$ where F_a is the allowance multiplier for PF&D (number), and PF&D is the personal, fatigue, and delay allowance (percentage).

Standard productivity is the time required by a trained and motivated worker or workers to perform construction task while working at normal tempo.

$Hs = T_n \times F_a$ where Hs is the standard time for a construction task per unit of effort (hour).

The allowance for PF&D is 15%, which is an allowance multiplier of 1.176.

1.9.2   Field data report

The field report is collected in the field for similar work, and a spreadsheet is devised for the data. The report is used for time control and to find the number of man-hours for a task. The spreadsheet prepares the data for statistical analysis. The estimator determines the productivity rate, and the rate is used for future cost analysis and estimating similar scopes of work.

1.9.3 Table—spreadsheet for field data report

Project: Combined cycle, install HRSG RP piping and supports
   Foreman: John Smith Date:
   Craft: PF

Cost code	Phase code description	QTY	MH
	RP piping and supports		
xxxxxx	RP-50 RH desuperheater spray water	65	116.2
xxxxxx	RP-51 reheater 1 to 2 drain	40	66.8
xxxxxx	1.5″ RP-50	60	105.4
xxxxxx	CRH boot drain	20	35.5
xxxxxx	HRH boot drain	20	35.5
xxxxxx	Preheater drains	180	244.7

1.10 Detailed estimate

The detailed estimate is developed by combining the unit method and historical man-hours into the unit-quantity model to determine the quantities and cost to prepare an accurate and timely estimate. The unit-quantity model has the advantage that the scope and quantity differences can be identified and the impacts estimated.

1.11 Unit method

The unit method uses historical man-hours and is defined as the mean, found as

$$x \text{ bar} = x_1 + x_2 + \cdots + x_n = \sum \frac{x}{n}$$

The unit method is the average man-hour per quantity n_i

$$MH_a = \frac{\sum MH\mu}{\sum n_i}$$

where MH_a is the average unit man-hour per quantity n_i; $MH\mu$ is the quantity n_i (unit man-hour); μ is the task 1, 2, k from quantity takeoff associated with construction work; and n_i is the task takeoff quantity i, in dimensional units.

1.11.1 Table—calculation HRSG RP piping and supports average man-hour (MH$_a$)

HRSG RP piping and supports	Qty n_i, lf	MHμ
RP-50 RH desuperheater spray water	65	116.2

(Continued)

(Continued)

HRSG RP piping and supports	Qty n_i, lf	MHμ
RP-51 reheater 1 to 2 drain	40	66.8
1.5″ RP-50	60	105.4
CRH boot drain	20	35.5
HRH boot drain	20	35.5
Preheater drains	180	244.7
	$\sum = 385.0$	$\sum = 604.1$

HRSG, Heat recovery steam generator.

$$\frac{\sum MH\mu}{\sum n_i} = 1.57$$

$$MH_a = 1.57 \, MH/LF$$

- The unit method is the slope b where a = 0 or y = 0 + bx; MH_a is a linear multiplier.
- When using MH_a, the unit method can either overestimate or underestimate MH_a.
- The data is plotted, and the values for y = a + bx are calculated using linear regression model (Fig. 1.1).

1.11.2 Regression model—unit method

The data is plotted, for y = a + bx, using linear regression:		y = 1.2975x + 17.428
	Quantity, n_i, lf	MHμ
HRSG RP piping and supports	x	y
RP-50 RH desuperheater spray water	65	116.2
RP-51 reheater 1 to 2 drain	40	66.8
1.5″ RP-50	60	105.4
CRH boot drain	20	35.5
HRH boot drain	20	35.5
Preheater drains	180	244.7

COVAR (R1, R2)	3875.47
VARP (R2)	2986.81
SLOPE (R1, R2)	1.2975
INTERCEPT (R1, R2)	17.428

HRSG, Heat recovery steam generator.

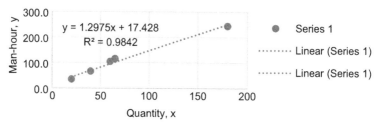

FIGURE 1.1 Unit method $MH_a = \Sigma MH\mu/n_i$.

1.11.3 Correlation

- Measures the strength and direction of a linear relationship between two variables.
- The value of r is such that $(-1 \leq r \leq +1)$. The $+$ and $-$ signs are used for positive and negative linear correlations, respectively.
- Positive correlation: If x and y have a strong positive linear correlation, r is close to $+1$. An r value exactly $+1$ indicates a perfect positive fit. Positive values indicate a relationship between X and Y variables such that as values for X increase, values for Y also increase.
- A perfect correlation of $+$ or -1 occurs only when all the data points lie exactly on a straight line. If $r = +1$, the slope line is positive. If $r = -1$, the slope of this line is negative.
- A correlation *greater than 0.8* is generally described as strong, whereas a correlation *less than 0.5* is generally described as weak.

Excel function—CORREL (R1, R2) = correlation coefficient of data in arrays R1 and R2
CORREL (R1, R2) = correlation coefficient = 0.9986
CORREL (R1, R2)^2 = coefficient determination = 0.9842
The coefficient of determination is $R^2 = 0.9842$ and the correlation coefficient, R = 0.9987 is a strong indication of correlation. A percentage of 99.8 of the total variation on Y can be explained by the linear relationship between X and Y (described by the regression equation; Y = 1.2975X + 17.428). The relationship between X and Y variables is such that as X increases, Y increases.

1.12 Unit-quantity model

Simplify the unit method and solve for $MH\mu$:
$$\sum MH\mu / \sum n_i = \sum MH_a.$$ Therefore $MH\mu = \sum n_i(MH_a)$ (*unit-quantity model*)
The model starts with the quantity takeoff arranged in the erection sequence required to assemble and install the construction task. The estimator selects the task description by defining the work scope for field

construction work. Each task is related to and performed by direct craft and divided into one or more subsystems that are further divided into assemblies made up of construction line items.

The unit-quantity model is given by:

$$MH\mu = \sum n_i(MH_a)$$

where $MH\mu$ is the quantity n_i (man-hours for field construction work), n_i is the task takeoff quantity i (dimensional units), MH_a is the average unit man-hour per quantity n_i, and μ is the task 1, 2, k from quantity takeoff associated with construction work.

The n_i quantity is the takeoff for the field scope of work.

MH_a man-hours are determined from field construction man-hour tables.

The unit man-hours were determined from historical data, and they correspond to the labor productivity necessary to install the civil, structural, and mechanical work in an industrial construction facility.

The unit-quantity model allows a final cross-check of actual man-hours to estimated man-hours.

$$Factor:MH\mu = n_i(MH_a)(fx)$$

where fx is the factor percent for productivity loss and alloy weld factor $x = 1, 2, p$.

1.12.1   Table—illustration of unit-quantity model

HRSG RP piping and supports

Scope of work	Quantity n_i	Unit	MH_a	Unit MH $MH\mu$
RP-50 RH desuperheater spray water	65	lf	1.57	102.05
RP-51 reheater 1 to 2 drain	40	lf	1.57	62.8
1.5″ RP-50	60	lf	1.57	94.2
CRH boot drain	20	lf	1.57	31.4
HRH boot drain	20	lf	1.57	31.4
Preheater drains	180	lf	1.57	282.6
$MH\mu = \sum n_i(MH_a)$				*604.5*

1.12.2 Regression model—unit-quantity model verification

The data is plotted, for $MH\mu = \sum n_i(MH_a)$, using linear regression (Fig. 1.2).

	Quantity, n_i, lf x	$MH\mu$ y
RP-50 RH desuperheater spray water	65	102
RP-51 reheater 1 to 2 drain	40	62.8

(Continued)

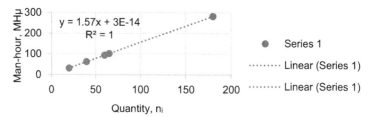

FIGURE 1.2 Unit-quantity model $MH\mu = \Sigma n_i(MH_a)$.

(Continued)

	Quantity, n_i, If x	$MH\mu$ y
1.5″ RP-50	60	94.2
CRH boot drain	20	31.4
HRH boot drain	20	31.4
Preheater drains	180	283
COVAR (R1, R2)	4689.28	
VARP (R2)	2986.81	
SLOPE (R1, R2)	1.5700	
INTERCEPT (R1, R2)	0.000	

The coefficient of determination, R^2, is exactly $+1$ and indicates a positive fit. All data points lie exactly on the straight line. The relationship between X and Y variables is such that as X increases, Y also increases.

The unit-quantity model improves the accuracy of the historical data.

1.13 Excel estimate spreadsheet for the unit quantity model

Estimate sheet—sample form for excel spreadsheet estimate		Historical		Estimate	
Description	MH	Qty	Unit	Qty	PF
Facility scope					0.0
Scope	0	0	LF	0.0	0.0
Scope	0	0	LF	0.0	0.0
Scope	0	0	LF	0.0	0.0

The Excel spreadsheet based on the unit-quantity model to estimate industrial construction has many advantages over other estimating methods:

- The scope of work is defined, and the erection sequence is based on the duplication of design.
- The unit-quantity model is a balanced unit estimate.
- The unit-quantity model is a balanced bid strategy.

- Standard unit of construction work is completed before construction is started.
- A detailed work estimate is the most accurate and timely work estimate.
- Using the duplication of design to define scope reduces the risk of lack of information.
- Planning occurs when the unit-quantity model is used.

1.14   Elements of construction work

- The detailed estimate is accurate, reliable, consistent, verifiable, and timely.
- Estimate consists of direct craft labor, material, equipment, and subcontract material.
- A chart of elements is included in a construction-work estimate:

Cost construction work

Subcontract material
Equipment
Material   Direct cost
Direct labor

1.15   Computer-aided cost estimation

Computerized estimating is a standard tool in the construction industry, which:

- reduces the time you spend in estimating,
- uses functions and formulas to ensure accurate estimates,
- shows step-by-step how to set up:
- estimate spreadsheets
- formulas for construction estimating
- detail sheets
- complete estimates in less time
- improve estimate accuracy
- organize the following:
- spreadsheet estimate
- summary sheet
- detailed workup sheet
- formulas that will automatically calculate man-hours and cost
- sheets containing detail and databases
- store and analyze data

By comparison, construction-estimating spreadsheets, given historical data from the previous similar work task, can reliably calculate man-hours, material, and labor by category and rapidly produce both summary and detailed man-hour and cost estimates. Manual takeoff with too much details introduces the potential for human error and inaccuracies. Using the unit-quantity model, takeoff counts, and measurements can be generated from the model. Therefore data is always consistent with the design. An automation of the tedious task of quantifying allows the estimator to focus on identifying construction assemblies, cost, and factoring risk.

Chapter 2

Construction material

2.1 Overview—introduction

The contractor must make a takeoff of the types and quantities of various materials to complete a construction project. This is to ensure a complete estimate of the cost and material requirements associated with the installation of the materials. This process is the material takeoff (MTO) that is an essential part of the estimation process. The MTO must be an accurate estimate of the cost and materials so the project can run smoother and the craft will have the materials they need. Accurate detailed material estimates are vital to the successful planning and execution of the project. Automation of the takeoff of materials is the quickest, least expensive, and most reliable.

2.2 Material estimate

The material estimate identifies all materials required to complete the project and is derived from the MTO to develop estimating worksheets for each construction activity. Items identified by the MTO are organized, tabulated, and summarized like material items by construction activity, resulting in a material summary list that becomes the material estimate for the project. Steps in creating the material estimate depend on the accuracy of the material estimate derived from the MTO, also known as the quantity takeoff. The takeoff is a count of how much material will be required for the project to complete the scope of work per the specifications provided in the bid package.

2.3 Material takeoff

The construction takeoff is a MTO.

- The MTO must list all the materials required to complete the project.
- The MTO should include any and all raw and prefabricated tools, such as piping, concrete, forms, and structural steel.
- The MTO needs to specify the type of material, such as type of pipe fittings, valves, pipe supports, structural steel, and concrete.
- The MTO is used to prepare the bill of materials (BOM) that is used to requisition and procure the necessary materials to complete the project.

Industrial Construction Estimating Manual. DOI: https://doi.org/10.1016/B978-0-12-823362-7.00002-8

- Cost of material = MTO × cost/quantity.
- The unit method is used for material pricing.

2.4 Estimate worksheet

The estimate worksheets illustrated in the sample MTOs are the basis for the MTO and material estimate. The construction estimate worksheet provides a quick, efficient, and comprehensive cost-effective form for estimating and developing the material cost sheets for construction projects.

2.5 Sample estimating worksheets for piping, structural, foundations, and vessels/towers

2.5.1 Power plant piping material takeoff

Project: Combined cycle power plant

	Start-up/ shutdown (LCL) MTO				
Revision:	X			Date: x/xx/ xxxx	
					Total
Description	Size	Quantity	Unit	$/Unit	$
A106B std smls pipe	6	602	lf	0.00	0.00
3000# A105 Sol	6 × 1	5	ea.	0.00	0.00
3000# A105 Sol	6 × 2	2	ea.	0.00	0.00
3000# A105 Sol	6 × .75	2	ea.	0.00	0.00
90-degree Elbow LR std A234WPB BW	6	17	ea.	0.00	0.00
45-degree Elbow LR std A234WPB BW	6	2	ea.	0.00	0.00
Straight Tee std Type Strainer	6	2	ea.	0.00	0.00
Ecco Red std A234WPB BW	6 × 4	1	ea.	0.00	0.00
Ecc Red std A234WPB BW	6 × 3	2	ea.	0.00	0.00
Induction Bend R = 5 OD 1.5 degrees	6	1	ea.	0.00	0.00
Induction Bend R = 5 OD 2.9 degrees	6	1	ea.	0.00	0.00
150# RFWN Flange A105	6	5	ea.	0.00	0.00
150# Flat Ring Gasket Mineral Fiber	6	5	ea.	0.00	0.00
Stud Bolt A193 Hot Dip Galv w/2 nuts 4.25 "	0.75	40	ea.	0.00	0.00
A106B xs slms pipe	4	677	lf	0.00	0.00
3000# A105 Sol	4 × 1	5	ea.	0.00	0.00
3000# A105 Sol	4 × 0.75	3	ea.	0.00	0.00
45-degree Elbow LR std A234WPB BW	4	15	ea.	0.00	0.00
90-degree Elbow LR std A234WPB BW	4	20	ea.	0.00	0.00
Tee A234WPB std BW	4	1	ea.	0.00	0.00

(Continued)

(Continued)

	Start-up/ shutdown (LCL) MTO				
Revision:	X			Date: x/xx/ xxxx	
					Total
Conc. Red std A234WPB BW	4 × 2	1	ea.	0.00	0.00
150# RFWN Flange A105	4	3	ea.	0.00	0.00
150# Flat Ring Gasket Mineral Fiber	4	3	ea.	0.00	0.00
Gate Valve	4	4	ea.	0.00	0.00
150# RFWN Flange A105	3	3	ea.	0.00	0.00
150# Flat Ring Gasket Mineral Fiber	3	3	ea.	0.00	0.00
90-degree Elbow SR std A234WPB BW	3	2	ea.	0.00	0.00
Expansion joint	3	2	ea.	0.00	0.00
Stud Bolt A193 Hot Dip Galv w/2 nuts 4.0″	5/8	48	ea.	0.00	0.00
A106B xs smls pipe	2	2	lf	0.00	0.00
A106B xs smls pipe	1	43	lf	0.00	0.00
90-degree Pipe bend 50 xs	1	8	ea.	0.00	0.00
Globe valve	1	5	ea.	0.00	0.00
A106B xs sms pipe	2	22	lf	0.00	0.00
Tee 3000# A105 SW	2	1	ea.	0.00	0.00
90 Ell 3000# A105 SW	2	5	ea.	0.00	0.00
A106B xs smls pipe	1	6	lf	0.00	0.00
Pipe bend 50 90-degree xs	1	2	ea.	0.00	0.00
A106B xs sms pipe	1.25	16	lf	0.00	0.00

2.5.2   Power plant structural steel quantity material takeoff

Project: Combined cycle power plant

	Revision:	X		Date: xx/ xx/xxxx	
			Steel weight	Total steel weight	
Equipment description	Component description	Quantity	Tons	Tons	
HRSG utility steel bridge		**186**	**204.6**		
	Light— 0−19 lb/ft.	Lot	10	11	
	Medium— 20−39 lb/ft.	Lot	30	33	
		Lot	96	105.6	

(*Continued*)

(Continued)

Equipment description	Component description	Revision: X	Quantity	Date: xx/ xx/xxxx Steel weight Tons	Total steel weight Tons
	Heavy— 40–79 lb/ft. X heavy— 80–120 lb/ft.	Lot		50	55
STG utility bridge steel					**187**
	Light— 0–19 lb/ft.	Lot		7.6	8.4
	Medium— 20–39 lb/ft.	Lot		25.4	27.9
	Heavy— 40–79 lb/ft.	Lot		90	99
	X heavy— 80–120 lb/ft.	Lot		47	51.7
STG utility bridge steel inside enclosure					**37.4**
	Light— 0–19 lb/ft.	Lot		1.5	1.7
	Medium— 20–39 lb/ft.	Lot		5.1	5.6
	Heavy— 40–79 lb/ft.	Lot		18	19.8
	X heavy— 80–120 lb/ft.	Lot		9.4	10.3
GSU transformer access platforms					**13.5**
	Light and medium	1		12.3	13.5
Stair tower and ladders					**14.2**
	Light and medium	3		3.2	10.4
	Stair treads	300			2.3
	Ladders	200			1.5
Iso phase support steel					**33**
	Light and medium	2		15	33
Dead-end structure					**33**
	Light and medium	3		10	33
230 kV switchyard structure including bus supports					**16.5**

(Continued)

(Continued)

	Revision:	X		Date: xx/xx/xxxx
			Steel weight	Total steel weight
Equipment description	Component description	Quantity	Tons	Tons
	Light and medium	1	15	16.5
Transmission line poles				**6**
	Light and medium	6	1	6
Grating, handrail, and toe plate				**17.1**
	2″ serrated floor grating	11,200		8.7
	1.5″ serrated grating	4100		3.1
	Handrail and toe plate	2100		5.3

2.5.3 Power plant foundations quantity material takeoff

Project: Combined cycle power plant

Description	Component description	Quantity	Concrete volume (CY)	Total concrete volume (CY)	Total rebar (ton)
Boiler-related foundations					
HRSG stack and heat recovery steam generator	50′ × 200′ × 5′ thk	1	1852	1944.6	107
HRSG RH blowdown tank and sump foundation	16′ × 12′ concrete	1	60.4	60.4	3.3
HRSG HP blowdown tank foundation	10′ × 10′ × 2.5′ thk	1	9.3	9.3	0.5
CEMS building	12′ × 12′ × 2.5′ thk	1	13.4	14.7	0.8
Air skid foundation	19′ × 7′ × 2′ thk	1	12.3	14.1	0.8

(*Continued*)

(Continued)

Description	Component description	Quantity	Concrete volume (CY)	Total concrete volume (CY)	Total rebar (ton)
Steam Cycle Chem Feed Area Fdn	38' × 15' × 2' thk	1	46.1	50.7	2.8
Boiler feedwater pump	22.75' × 11.25' thk	2	29.7	65.3	3.6
Auxiliary boiler foundation	30' × 20' × 3.0' thk	1	66.7	73.4	4.0
CTG and CTG-related foundations					
Combustion turbine and CT generator	53' × 103'	1	1230.0	1291.5	107.0
Gen. circuit breaker foundation	25' × 10.5' × 2.5' thk	1	38.2	76.4	4.2
Rotor air cooler foundation	33' × 8' × 2' thk	1	19.6	21.6	1.2
Excitation skid and transformer	22.3' × 13' × 2' thk	1	21.5	23.7	1.3
VT and surge cubicle foundation	9.5' × 5' × 2' thk	1	6.0	6.6	0.4
Sample panel	24' × 8' × 2' thk	1	14.2	15.6	0.9
Compressor wash skid	18' × 9' × 2' thk	1	12.0	13.2	0.7
GT water wash tank	16' × 6' × 1.5' thk	1	5.3	5.8	0.3
STG and STG-related foundations					
Steam turbine and steam turbine generator	116' × 63' × 6' thk	1	2510.0	2635.5	145.0
Closed cooling water heat exchangers	13' × 9.75' × 2.5' thk	2	11.7	26.9	1.5
Closed cooling water pumps	11.25' × 5' × 3' thk	2	6.3	14.4	0.8
Cooling Tower chemical feed area	36.5' × 30' w/ walls	1	102.0	117.3	6.5
Circulating water pump station and cooling tower	44' × 27' and 72' × 120'	1	1350.0	1552.5	85.4
Sulfuric acid, coagulant tank,	51' × 17.5' × 2'	1	79.0	90.9	5.0

(Continued)

(Continued)

Description	Component description	Quantity	Concrete volume (CY)	Total concrete volume (CY)	Total rebar (ton)
and pump foundations					
Condenser polishing unit	14′ × 10′ × 2.5′ thk	1	13.0	15.0	0.8
Condensate extraction pumps	15′ × 7′ × 2.5′ thk	1	9.7	11.2	0.6
Condenser vacuum pumps (2)	16′ × 13′ × 2.5′ thk	1	19.3	22.2	1.2

2.5.4   Compressor station piping material takeoff

Project: Compressor station

Preliminary mechanical MTO list

Revision:	X			Date:
Quantity	Size	Item	Sch	Description
100′	30″	Pipe		Pipe, steel, 30.00″ OD × 0.625″ wall, DSAW
40′	12″	Pipe	XS	Pipe, steel, 12.00″ OD × 0.500″ wall, DSAW
2	30″	90-degree ELL.		Elbow, pipe, W E, 90-degree, L R, 0.625″ WT, GR Y60
5	12″	90-degree ELL.	XS	Elbow, pipe, WE, 90-degree, L R, 0.500″ WT, GR Y60
3	30″	45-degree ELL.		Elbow, pipe, weld ends, 45-degree, long radius, 0.625″
1	30″	Tee		Tee, pipe, welding, straight, 0.625″ WT, Y60, E B
2	30″ × 16″	Red tee		Tee, pipe, welding, Red, 30″ × 16″, 0.625″ WT, Y60, E B
10	30″	Flange	600#	Flange, welding neck, 1/4″ R F, 600#
11	30″	Gasket	600#	Gasket, G-10 glass epoxy, ANSI 600 LB Class
4	12″	Gasket	600#	Gasket, G-10 glass epoxy, ANSI 600 LB Class
1	16″ × 12″	Reducer		Red, weld ends, concentric, O.500″ WT, GR Y60,
1	16″ × 8″	Reducer		Red, weld ends, concentric, O.500″ WT, GR Y60,

(Continued)

(Continued)

Preliminary mechanical MTO list

Revision:	X			Date:
Quantity	Size	Item	Sch	Description
1	12″ × 8″	Reducer		Red, weld ends, concentric,O.500″ WT, GR Y60,
100′	42″	Pipe		Pipe, steel, 42.00″ OD × 0.875″ wall, API-5L, × 60, DSAW
390′	36″	Pipe		Pipe, steel, 36.00″ OD × 0.750″ wall, API-5L, × 60, DSAW
100′	30″	Pipe		Pipe, steel, 30.00″ OD × 0.625″ wall, API-5L, × 60, DSAW
16′	16″	Pipe		Pipe, steel, 16.00″ OD × 0.500″ wall, API-5L, × 60, DSAW
40′	12″	Pipe	XS	Pipe, steel, 12.00″ OD × 0.500″ wall, API-5L, × 60, DSAW
2	30″	90-degree ELL.		Elbow, Pipe, W E, 90 degrees, L R, 0.625″ WT, GR Y60
1	16″	90-degree ELL.		Elbow, Pipe, W E, 90 degrees, L R, 0.500″ WT, GR Y60
5	12″	90-degree ELL.	XS	Elbow, Pipe, W E, 90 degrees, L R, 0.500″ WT, GR Y60
6	30″	45-degree ELL.		Elbow, Pipe, W E, 45 degrees, L R, 0.625″ WT, GR Y60
1	42″	Tee		Tee, Pipe, welding, straight, 0.875″ WT, Y60
6	30″	Tee		Tee, Pipe, welding, straight, 0.625″ WT, Y60
3	30″ × 16″	Red Tee		Tee, Pipe, welding, reducing, 30″ × 16″, 0.625″ WT, Y60
2	36″	Flange	600#	Flange, welding neck, 1/4″ raised face, Grade F60
10	30″	Flange	600#	Flange, welding neck, 1/4″ raised face, Grade F60
2	36″	Blind flange	600#	Flange, blind, 1/4″ raised face, Grade F6
2	36″	Gasket	600#	Gasket, G-10 glass epoxy, ANSI 600 LB Class
11	30″	Gasket	600#	Gasket, G-10 glass epoxy, ANSI 600 LB Class
4	12″	Gasket	600#	Gasket, G-10 glass epoxy, ANSI 600 LB Class
1	42″ × 30″	Reducer		Reducer, weld ends, concentric, O.875″ WT, GR Y60,
1	42″ × 22″	Reducer		Reducer, weld ends, concentric, O.875″ WT, GR Y60

(Continued)

(Continued)

Preliminary mechanical MTO list

Revision:	X			Date:

Quantity	Size	Item	Sch	Description
1	22″ × 16″	Reducer		Reducer, weld ends, concentric, O.500″ WT, GR Y60
1	16″ × 12″	Reducer		Reducer, weld ends, concentric, O.500″ WT
3	16″ × 8″	Reducer		Reducer, weld ends, concentric, O.500″ WT, GR Y
1	12″ × 8″	Reducer		Reducer, weld ends, concentric, O.500″ WT, GR Y60
100′	30″	Pipe		Pipe, steel, 30.00″ OD × 0.625″ wall, API-5L, × 60, DSAW
40′	12″	Pipe	XS	Pipe, steel, 12.00″ OD × 0.500″ wall, API-5L, × 60, DSA
2	30″	90-degree ELL.		Elbow, Pipe, W E, 90-degree, L R, 0.625″ WT, GR Y60
5	12″	90-degree ELL.	XS	Elbow, Pipe, W E, 90-degree, L R, 0.500″ WT, GR Y60
3	30″	45-degree ELL.		Elbow, Pipe, W E, 45-degree, L R, 0.625″ WT, GR Y60
1	30″	Tee		Tee, Pipe, welding, straight, 0.625″ WT, Y60 EB
2	30″ × 16″	Red tee		Tee, Pipe, welding, Red, 30″ × 16″, 0.625″ WT, Y60, E B
10	30″	Flange	600#	Flange, welding neck, 1/4″ R F, 600#
11	30″	Gasket	600#	Gasket, G-10 glass epoxy, ANSI 600 LB Class
4	12″	Gasket	600#	Gasket, G-10 glass epoxy, ANSI 600 LB Class
1	16″ × 12″	Reducer		Reducer, W E, concentric, O.500″ WT, GR Y60
1	16″ × 8″	Reducer		Reducer, W E, concentric, O.500″ WT, GR Y60
1	12″ × 8″	Reducer		Reducer, W E, concentric, O.500″ WT, GR Y60
12′	36″	Pipe		Pipe, steel, 36.00″ OD × 0.750″ wall, API-5L, × 60

2.5.5 Vessels/columns material takeoff

	Diameter	H or L	Volume	Weight
Vessels/columns	FT-IN	FT-IN	FT^3	Ton
Scope				
Solvent absorber	8′−6″	69′−0″	3915	319.7
Unload, handle, haul up to 2000′, rig, set and align, make up foundation AB				
Install platforms and ladders				
Remove and replace manway cover (24″ 300# Removable-Davit)				
Install double downflow valve trays (16 trays)				
Install demisting pads (single grid support, pad, grid top)				
Vortex breaker				
Packing (pall rings)				
Rich solvent flash drum	7′−0″	24′−0″	924	14.9
Unload, handle, haul up to 2000′, rig, set and align, make up foundation AB				
Install platforms and ladders				
Remove and replace manway cover (24″ 300# Hinged)				
Inlet box				
Vortex breaker				
Install demisting pads (single grid support, pad, grid top)				
Solvent regenerator	8′−0″	72′−6″	3644	39.0
Unload, handle, haul up to 2000′, rig, set and align, and make up foundation AB				
Install platforms and ladders				
Remove and replace manway cover (24″ 300# Removable-Davit)				
Install double downflow valve trays (12 trays)				
Install demisting pads (single grid support, pad, grid bottom)				
Vortex breaker				
Packing (pall rings)				
Solvent regenerator reflux drum	5′−0″	15′−0″	295	2.6
Unload, handle, haul up to 2000′, rig, set and align, make up foundation AB				
Install platforms and ladders				
Remove and replace manway cover (24″ 300# Hinged)				
Install demisting pads (single grid support, pad, grid top)				
Vortex breaker				

(Continued)

(Continued)

	Diameter	H or L	Volume	Weight
Vessels/columns	FT-IN	FT-IN	FT^3	Ton
Solvent sump drum	$4'-0''$	$12'-0''$	151	2.2

Solvent sump drum
Unload, handle, haul up to 2000', rig, set and align, make up foundation AB
Install platforms and ladders
Remove and replace manway cover (24″ 300# Hinged)
Vortex breaker
Internal pipe

2.6 Combined cycle power plant material takeoff

HRSG vendor piping MTO

2.6.1 LB code piping sheet 1

				LB			LB	LB		
Line no.	Material	Size	Sch/The	Pipe	BW	PWHT	Valve	Boltup	Instrument	PS
HP piping and supports—ASME Section 1										
HP-03 Econ 1 to HP Econ 2	SA-106-B	8	140/0.906	25						
HP-03 Econ 1 to HP Econ 2	SA-106-B	6	160/0.906	42	9	9			9	
HP-04 Econ 2 to HP Steam Drum	SA-106-B	8	140/0.812	96	4	4			6	2
HP-04 Econ 2 to HP Steam Drum	SA-106-C	6	160/0.906		3	3				
HP-09 Steam Drum to HP SH 1	SA-106-B	6	160/0.906	24	4	4			4	1
HP-10 Steam Drum to HP SH 1	SA-106-B	6	160/0.906	23	3	3			4	1
HP-11 Steam Drum to HP SH 1	SA-106-B	6	160/0.906	23	4	4			4	1
HP-12 Steam Drum to HP SH 1	SA-106-B	6	160/0.906	23	3	3			4	1
HP-13 Steam Drum to HP SH 1	SA-106-B	6	160/0.906	23	4	4			4	1
HP-14 Steam Drum to HP SH 1	SA-106-B	6	160/0.906	23	3	3			4	1
HP-15 SHTR 1 to HP SHTR 2	SA-335-P91	10	1.25	68	4	4			5	2
HP-15 SHTR 1 to HP SHTR 2	SA-335-P91	8	1	3	2	2				
HP-15 SHTR 1 to HP SHTR 2	SA-335-P91	3					1			
HP-16 SHTR 1 to HP SHTR 2	SA-335-P91	10	1.25	68	4	4			5	2

(*Continued*)

(Continued)

Line no.	Material	Size	Sch/The	Pipe (LB)	BW	PWHT	Valve (LB)	Boltup (LB)	Instrument	PS
HP-16 SHTR 1 to HP SHTR 2	SA-335-P91	8	1	3	2	2				
HP-16 SHTR 1 to HP SHTR 2	SA-335-P91	3					1			
HP-17 SHTR 1 to HP SHTR 2	SA-335-P91	10	1.25	68	4	4			5	2
HP-17 SHTR 1 to HP SHTR 2	SA-335-P91	8	1	3	2	2				
HP-17 SHTR 1 to HP SHTR 2	SA-335-P91	3					1			
IP Piping and Supports—ASME Section 1										
IP-03 Econ to IP Steam Drum	SA-106-B	4	40	53	9.0		2		3	10
IP-03 Econ to IP Steam Drum	SA-106-B	3	40	6	4.0		2			2
IP-06 Drum to IP Superheater 1	SA-106-B	10	40	42	2.0			4		
IP-06 Drum to IP Superheater 1	SA-106-B	6	40	5	6.0					
IP-09 SH1 to IP SH 2	SA-106-B	10	40	30	2.0			2		
LP Piping and Supports—ASME Section 1										
LP-10 Steam Drum to LP SHTR 1	SA-106-B	14	std	26				4		
LP-10 Steam Drum to LP SHTR 1	SA-106-B	10	40	34	4.0					
LP-10 Steam Drum to LP SHTR 1	SA-106-B	8	40	12	3.0					
LP-14 SHTR 1 to LP SHTR 2	SA-106-B	10	40	32	2.0			2		
LP-15 SHTR 1 to LP SHTR 2	SA-106-B	10	40	30	2.0			2		
RP Piping and Supports—ASME Section 1										
RP-03 Reheater 1 to Reheater 2	SA-335-P91	24	80/1.218	137	4.0	4			2	2
RP-03 Reheater 1 to Reheater 2	SA-335-P91	24	80/0.843		5.0	5				
RP-03 Reheater 1 to Reheater 2		3					1			

2.6.2 LB code piping sheet 2

				LB		LB	LB		
Line no.	Material	Size	Sch/The	Pipe BW	PWHT	Valve	Boltup	Instrument	PS
HP piping and supports—ASME B31.1									
HP-01 Feedwater Inlet	SA-106-C	10	160/1.125	47	2.0 2			3	1
HP-01 Feedwater Inlet	SA-106-C	6	160/0.718	3	3.0				
HP-01 Feedwater Inlet	SA-106-C	10	1.5	57	5.0 5	1			3
HP-02-Steam Outlet	SA-335-P91	14	1.75	44	5.0 5	1		5	2
HP-02-Steam Outlet	SA-335-P91	10	1.25	26	6.0 4			3	
HP-18 SHP-PSV-104 to SHP-S1−002		6				1			
HP-19 Start-Up Vent Piping	SA-335-P91	8	120/0.718	22	2.0 2		2		
HP-20 Sparging Steam Header	SA-106-B	3	160/0.437	42	5.0	3			4
HP-21 SHP-PSV-112A to SHP-S1-001 (Drum)		6					2		
HP-22 SHP-PSV-112B to SHP-S1-001 (Drum)		6					2		
HP-23 Upper Drain Piping Coll Header	SA-106-B	8	40	1					
HP-23 Upper Drain Piping Coll Header	SA-106-B	3	40	66	1.0				
HP-24 Steam Drum Warming Connection	SA-106-B	4	160	16	6.0	1			
HP-25 SHP-PSV-104 Stack Piping	SA-106-B	14	40	29	1.0				
HP-25 SHP-PSV-104 Stack Piping	SA-106-B	16	40	1					
HP-26 SHP-PSV-112A Stack Piping	SA-106-B	18	40	1					
HP-26 SHP-PSV-112A Stack Piping	SA-106-B	16	40	12	2.0				
HP-27 SHP-PSV-112B Stack Piping	SA-106-B	18	1						
HP-27 SHP-PSV-112B Stack Piping	SA-106-B	16	12		2.0				
IP Piping and Supports—ASME B31.1									
IP-01 Feedwater Inlet	SA-106-B	4	40	136	10.0	1		2	9
IP-02-Steam Outlet	SA-106-B	10	40	84	2.0	2		3	5
IP-02-Steam Outlet	SA-106-B	8	40	4	4.0			1	
IP-10 Pegging Steam	SA-106-B	16	std	7	3.0	1			
IP-10 Pegging Steam	SA-106-B	6	40	18	3.0	2			
IP-11 SIP-PSV-109 to SIP-SI-002 (Stem Out)		4				1	2		
IP-12 Start-Up to SIP-SI-005	SA-106-B	6	40	19	5.0	1			
IP-13 SIP-PSV-106 to SIP-SI-001 (Drum)		6				1	2		

(Continued)

(Continued)

Line no.	Material	Size	Sch/The	Pipe BW	LB PWHT	Valve	LB Boltup	LB Instrument	PS
IP-15 Sparging Steam Header	Sa-106-B	3	40	38	5.0	3			
IP-16 Intermittent Blow off	SA-106-B	2.5	80	84	6.0				6
IP-17 SIP-PSV-109 Stack Piping	SA-106-B	10	40	1		1			
IP-17 SIP-PSV-109 Stack Piping	SA-106-B	8	40	20	1.0				
IP-18 SIP-PSV-106 Stack Piping	SA-106-B	12	40	1					
IP-18 SIP-PSV-106 Stack Piping	SA-106-B	10	40	16	1.0				
IP-19 Flash Tank 1 to IP Drum	SA-106-B	2.5	80	18	2.0				
IP-20 F.S. 1 to Blowdown Tank	SA-106-B	6	80	1					
IP-21 BFW-PSV 200 Outlet Piping	SA-106-B	2.5	160	102	4.0		1		5
IP-21 BFW-PSV 200 Outlet Piping	SA-106-B	2.5	160	102	4.0		1		5

2.6.3 LB code piping sheet 3

Line no.	Material	Size	Sch/The	Pipe BW	LB PWHT	Valve	LB Boltup	LB Instrument	PS
LP Piping and Supports—ASME B31.1									
LP-01 Feedwater Inlet	SA-106-B	8	40	109	5.0	1	2	3	5
LP-01 Feedwater Inlet	SA-106-B	6	40		2.0				
LP-02 Steam Outlet	SA-106-B	14	std	65	6.0			6	4
LP-02 Steam Outlet	SA-106-B	10	40	23	8.0	2		2	
LP-02 Steam Outlet	SA-106-B	8	std	1					
LP-02 Steam Outlet		3					1		
LP-03 LP FW Heater Bypass	SA-106-B	8	40	3	2.0		2		
LP-04 FW Heater 1 to FW Heater 2	SA-106-B	6	40	5	2.0			1	
LP-05 FW Heater 1 to FW Heater 2	SA-106-B	6	40	5	2.0			1	
LP-06 FW Heater 1 to FW Heater 2	SA-106-B	6	40	5	2.0			1	
LP-07 FW Heater 2 to LP Steam Drum	SA-106-B	8	40	45	13.0	6		4	2
LP-07 FW Heater 2 to LP Steam Drum	SA-106-B	6	40	6	5.0	1			
LP-16 Boiler Feed Pump Recirc.	SA-106-B	6	40	49	3.0	1	1		

(Continued)

(Continued)

Line no.	Material	Size	Sch/The	Pipe BW LB	PWHT	Valve LB	Boltup LB	Instrument	PS
LP-16 Boiler Feed Pump Recirc.	SA-106-B	6	1	6					
LP-17 Boiler Feed Pump Recirc.	SA-106-B	6	40	43	3.0	1	1		
LP-17 Boiler Feed Pump Recirc.	SA-106-B	6	1	6					
LP-20 RAC to LP Steam Drum	SA-106-B	10	40	203	8.0				7
LP-21 SLP-PSV-102 to SLP-SI-002		6					2		
LP-22 Start-Up Vent to SLP-SI-003	SA-106-B	8	40	19	5.0				
LP-23 SLP-PSV-104A to SLP-SI-001 (Drum)		6					2		
LP-24 SLP-PSV-104B to SLP-SI-001 (Drum)		6					2		
LP-26 Sparging Steam Header	SA-106-B	3	40	38	5.0	3			1
LP-27 Intermittent Blow off	SA-106-B	2.5	80	91	8.0				5
LP-28 SLP-PSV-102 Stack Piping	SA-106-B	10	40	1					
LP-28 SLP-PSV-102 Stack Piping	SA-106-B	8	40	20	1.0				
LP-29 SLP-PSV-104A Stack Piping	SA-106-B	10	40	1					
LP-29 SLP-PSV-104A Stack Piping	SA-106-B	8	40	12	2.0				
LP-30 SLP-PSV-104B Stack Piping	SA-106-B	10	40	1					
LP-30 SLP-PSV-104B Stack Piping	SA-106-B	8	40	11	1.0				

RP Piping and Supports—ASME B31.1

Line no.	Material	Size	Sch/The	Pipe BW LB	PWHT	Valve LB	Boltup LB	Instrument	PS
RP-01 Reheater 1 Inlet	SA-106-B	24	60/1.218	60	2.0 2		2		3
RP-01 Reheater 1 Inlet	SA-106-B	16	60/0.843	14	12.0 12				
RP-01 Reheater 1 Inlet		6				2			
RP-02 Reheater 3 Outlet	SA-335-P91	24	60/1.218	50	2.0 2		4		
RP-02 Reheater 3 Outlet	SA-335-P91	16	60/0.843	24	6.0 6		3		
RP-02 Reheater 3 Outlet	SA-335-P91	10	40	1					
RP-02 Reheater 3 Outlet		4				1			
RP-09 SIP-PSV-108A to SIP-SI-003(CLD RH)		6				2			
RP-10 SIP-PSV-108B to SIP-SI-003(CLD RH)		6				2			
RP-11 SIP-PSV-105 to SIP-SI-004(RH Out)		6				2			
RP-12 Start-Up Vent	SA-335-P91	10	40	27	5.0 5	1			
RP-13 SIP-PSV-108A Stack Piping	SA-106-B	18	40	1					

(*Continued*)

(Continued)

Line no.	Material	Size	Sch/The	Pipe	LB BW	PWHT	LB Valve	Boltup	LB Instrument	PS
RP-13 SIP-PSV-108A Stack Piping	SA-106-B	16	40	19	1.0					
RP-13 SIP-PSV-108B Stack Piping	SA-106-B	20	40	1						
RP-13 SIP-PSV-108B Stack Piping	SA-106-B	18	40	16	1.0					
RP-15 SIP-PSV-105 Stack Piping	SA-106-B	14	40	1						
RP-15 SIP-PSV-105 Stack Piping	SA-106-B	12	40	20	1.0					
RP-03 Cooling Air Piping	SA-106-B	3	40	28	1.0			2		1

2.6.4 LB code piping sheet 4

Line no.	Material	Size	Sch/The	Pipe	LB BW	PWHT	LB Valve	Boltup	LB Instrument	PS
Silencer drain piping and supports										
14″-RP-03 PRVS-X703B to SIL-X704	SA-106-B	14	40	40	2.0		0	0		
10″-RP-05 PRVS-X703A to SIL-X704	SA-106-B	14	40	40	2.0					
10″-RP-10 PRVS-X704 to SIL-X705	SA-106-B	10	40	40	2.0					
12″-RP-08 MOV-X211/ 212 SIL-X706	SA-106-B	12	40	40	6.0		2			
10″-HP-13 PRVS-X802 to SIL-X803	SA-106-B	10	40	40	2.0					
12″-HP-17 PRVS-X801A to SIL-X801	SA-106-B	12	40	40	2.0					
8″-LP-13 PRVS-X602 to SIL-X603	SA-106-B	8	40	40	2.0					
6″-IP-10 PRVS-X702 to SIL-X703	SA-106-B	6	40	40	2.0					
4″-IP-08 MOV-X205/206 to SIL-X702	SA-106-B	4	40	40	6.0		2			
8″-IP-12 PRVS-701 to SIL-X701	SA-106-B	8	40	40	2.0					
12″-LP-11 MOV-X106/ 107 to SIL-X602	SA-106-B	12	40	40	6.0		2			
8″-LP-15 PRVS-X601A to SIL-X601	SA-106-B	8		40	2.0					
8″-LP-17 PRVS-X601B to SIL-X601	SA-106-B	8		40	2.0					
12″-HP-15 PRVS-X801B to SIL-X801	SA-106-B	12	40	40	2.0					
6″-IP-11 MOV-X310/ 311to SIL-X803	SA-106-B	6	40	40	7.0		2			
PSV Drip Pan Assembly	SA-106-B	6	40	140			13			

2.6.5   Small bore code piping sheet 5

Line no.	Material	Size	Sch/ Thk	SB Pipe	SW	SB Valve	ps	Instrument
HP piping and supports								
HP-51 Superheater 2 Drain	SA-106-B	2	80	80	41	3	4	
HP-50 Superheater 1 Drain	SA-106-B	2	80	80	34	3	4	
HP Economizer 1 Drains	SA-106-B	2	80	140	56	7	7	
HP Economizer 2 Drains	SA-106-B	2	80	220	66	11	11	
HP-18-A Sparging Header	SA-106-B	2	80	20	10	2	1	
HP-23-D manifold Drain	SA-106-B	2	80	20	8	1	1	
HP-52 Upper Blowdown Piping	SA-106-B	2	80	100	20		5	
HP Upper Drains	SA-106-B	2	80	100	20		5	
IP Piping and Supports								
IP continuous blowdown to continuous blowdown tank	SA-106-B	1.5	80	120	44	8	6	4
IP-50 SPHTR 1 to SPHTR 2 Drain	SA-106-B	2	80	80	18	2	4	
IP-13-A Sparging Steam Header Drain	SA-106-B	2	80	20	12	2	1	
IP Economizer Drains	SA-106-B	1	80	140	42	7	7	
2″-IP-51	SA-106-B	2	80	120	26	2	6	
IP-16-D Drain	SA-106-B	2	80	20	4	1	1	

(*Continued*)

(Continued)

Line no.	Material	Size	Sch/Thk	SB Pipe	SW	SB Valve	ps	Instrument
IP-51 Blowdown Piping	SA-106-B	2	80	100	20		5	
IP/RH Upper Drains	SA-106-B	2	80	100	20		5	
LP piping and supports								
LP-50 SPHTR 1 to SPHTR 2 Drain	SA-106-B	2	80	140	20	3	7	
LP-22-A Sparging Steam Drain	SA-106-B	2	80	20	12	2	1	
LP-22-C Drain	SA-106-B	2	80	20	14	3	1	
LP-26-A Blow off Tank Drain	SA-106-B	2	80	20	16		1	
LP-51 Upper Blowdown Piping	SA-106-B	2	80	100	20		5	
LP boot drain	SA-106-B	2	80	20	4		1	
LP upper drains	SA-106-B	2	80	100	20		5	
RP piping and supports								
RP-50 RH Desuperheater Spray water	SA-106-B	1.5	80	65	25	5	3	
RP-51 Reheater 1 to 2 Drain	SA-106-B	2	80	40	12	2	2	
1-1/2" RP-50	SA-106-B	1.5	80	60	19	7	3	
CRH boot drain	SA-106-B	2	80	20	4		1	
HRH boot drain	SA-106-B	2	80	20	4		1	
Preheater drains	SA-106-B	2	80	180	54	9	9	
Blowdown piping								
HD-D'-N Intermittent Blow off	SA-106-B	1.5	80	120	20	2	6	2

(Continued)

(Continued)

Line no.	Material	Size	Sch/ Thk	SB Pipe	SW	SB Valve	ps	Instrument
HD-01-M Cascading Blowdown to IP Drum	SA-106-B	1.5	80	120	64	11	6	4
ID-01-M Intermittent Blow off	SA-106-B	1.5	80	120	20	2	6	2
LD-01-N Intermittent Blow off	SA-106-B	1.5	80	120	20	2	6	2
HD-01-N Intermittent Blow off	SA-106-B	2	80	100	20		5	
LD-03-G BD Tank to BD Tank	SA-106-B	2	80	100	20		5	
Instrumentation								
Casing Instrumentation	SA-106-B	1.5	80	160	384	96		32

2.6.6 Risers and down comers sheet 6

Line no.	Material	Size	Sch/Thk	LB Pipe	BW	PWHT	PS
HRSG-Risers and Down comers HP Drum Down comer **Line no.**							
HP-07 Down comer	SA-106-B	26	2.5	192	6	6	4
HP-07 Down comer	SA-106-C	16	2	40	6	6	
HP-07 Down comer		4		10			
HP Evap to HP Drum Risers							
AHR11	SA-106-C	8	140	10	2	2	
AHR12	SA-106-C	8	140	10	2	2	
AHR13	SA-106-C	8	140	10	2	2	
BHR11	SA-106-C	8	140	10	2	2	
BHR12	SA-106-C	8	140	10	2	2	
BHR13	SA-106-C	8	140	10	2	2	

(*Continued*)

(Continued)

Line no.	Material	Size	Sch/Thk	LB Pipe	BW	PWHT	PS
CHR11	SA-106-C	8	140	10	2	2	
CHR12	SA-106-C	8	140	10	2	2	
CHR13	SA-106-C	8	140	10	2	2	
AHR21	SA-106-C	8	140	10	2	2	
AHR22	SA-106-C	8	140	10	2	2	
BHR21	SA-106-C	8	140	10	2	2	
BHR22	SA-106-C	8	140	10	2	2	
CHR21	SA-106-C	8	140	10	2	2	
CHR22	SA-106-C	8	140	10	2	2	
AHR31	SA-106-C	8	140	10	2	2	
AHR32	SA-106-C	8	140	10	2	2	
AHR33	SA-106-C	8	140	10	2	2	
BHR31	SA-106-C	8	140	10	2	2	
BHR32	SA-106-C	8	140	10	2	2	
IP Drum							
IP-05 Down comer	SA-106-B	10	40	114	3.0		2
IP-05 Down comer	SA-106-B	5	40	9	3.0		
IP-05 Down comer		4		2			
IP Evap to IP							
Drum Risers							
AIR11	SA-106-B	10	40	6	2		
AIR12	SA-106-B	10	40	6	2		
BIR11	SA-106-B	10	40	6	2		
BIR12	SA-106-B	10	40	6	2		
CIR21	SA-106-B	10	40	6	2		
CIR22	SA-106-B	10	40	6	2		
LP Drum							
Down comer							
LP-09 Down comer	SA-106-B	10	40	110	3.0		2
LP-09 Down comer	SA-106-B	6	40	20	3.0		
LP-09 Down comer		4		2			
LP Evap to LP							
Drum Risers							
ALR11	SA-106-B	12	80	10	2		
ALR12	SA-106-B	12	80	10	2		
BLR11	SA-106-B	12	80	10	2		
BLR12	SA-106-B	12	80	10	2		
CLR11	SA-106-B	12	80	10	2		
CLR12	SA-106-B		80	10	2		

2.6.7   Heat recovery steam generator—field trim piping sheet 7

	Material	Size	Sch/ Thk	SB Pipe	SW	SB Valve	Instrument
HP remote steam drum							
HD-01-K PI-X305	SA-106-B	1	80	20	4		2
HD-01-F LG-X301 w/ LT-X301C (Top Conn) HD-01-G LG-X301 w/LT-X301C (Bolt Conn) HD-01-H LG-X301 (Drain)	SA-106-B	1	80	20	4		6
HD-01-A LI-X301 w/ LT-X310A (Top Conn) HD-01-B LI-X301 w/ LT-X301A (Bolt Conn) HD-01-C LI-X301 (Drain)	SA-106-B	1	80	20	4		3
HD-01-D LT-X301B w/PT-X305B (Top Conn) HD-01-E LT-X301B w/PT-X305B (Bolt Conn)	SA-106-B	1	80	20	4		2
IP remote steam drum							
ID-01-F LG-X201 w/ LT-X201C (Top Conn) ID-01-G LG-X201 w/ LT-X201C (Bolt Conn) ID-01-H LG-X201 (Drain)	SA-106-B	1	80	20	50		6
ID-01-A LI-X201 w/ LT-X210A (Top Conn) ID-01-B LI-X201 w/ LT-X201A (Bolt Conn) ID-01-C LI-X201 (Drain)	SA-106-B	1	80	20	42	10	3
ID-01-D LT-X201B w/ PT-X203B (Top Conn) ID-01-E LT-X201B w/ PT-X203B (Bolt Conn)	SA-106-B	1	80	20	16	4	2
ID-01-K PI-X203	SA-106-B	1	80	20	12	4	2

(Continued)

(Continued)

	Material	Size	Sch/Thk	SB Pipe	SW	SB Valve	Instrument
LP remote steam drum							
LD-01-F LG-X101 w/ LT-X101C (Top Conn) LD-01-G LG-1201 w/ LT-X101C (Bolt Conn) LD-01-H LG-X101 (Drain)	SA-106-B	1	80	20	50	14	6
LD-01-A LI-X101 w/ LT-X110A (Top Conn) LD-01-B LI-X101 w/ LT-X101A (Bolt Conn) LD-01-C LI-X101 (Drain)	SA-106-B	1	80	20	42	10	3
LD-01-D LT-X101B w/PT-X103B (Top Conn) LD-01-E LT-X101B w/ PT-X103B (Bolt Conn)	SA-106-B	1	80	20	16	4	2
LD-01-N PI-X106	SA-106-B	1	80	20	12	4	2
Instrumentation							
IP Instrumentation	SA-106-B	0.75	80	420	240	70	27
HP Instrumentation	SA-106-B	0.75	80	420	260	78	21
HP Instrumentation		0.75	80	400	220	66	25
Atmospheric blow off tank							
LD-04-A LG-X103 (Top Conn) LD-04-B LG-X103 (Bolt Conn)	SA-106-B	1	80	20	23	8	6
PI-X114	SA-106-B	1	80	20	12	4	2
TI-X105	SA-106-B	1	80	5	4		1
Flash separator							
LD-04-A LG-X103 (Top Conn) LD-04-B LG-X103 (Bolt Conn)	SA-106-B	1	80	20	23	8	6

(*Continued*)

(Continued)

Material	Size	Sch/Thk	SB Pipe	SW	SB Valve	Instrument
PI-X114	SA-106-B	1	80	20	12	4	2
TI-X105	SA-106-B	1	80	5	4		1

2.6.8 SP-01 AIG piping sheet 8

	Material	Size	Sch/Thk	LB	BW	BU	PS
SP-01 AIG Manifold Connecting Piping Inner connect pipe	SA-106-B	10	40	73	3.0	3	2
SP-02 AIG Piping	SA-106-B	3	40	28	2.0	2	2
SP-03 AIG Piping	SA-106-B	3	40	21	2.0	2	2
SP-04 AIG Piping	SA-106-B	3	40	15	2.0	2	1
SP-05 AIG Piping	SA-106-B	3	40	10	1.0	2	1
SP-06 AIG Piping	SA-106-B	3	40	30	2.0	2	2
SP-07 AIG Piping	SA-106-B	3	40	37	2.0	2	3
SP-08 AIG Piping	SA-106-B	3	40	43	2.0	2	3
SP-09 AIG Piping	SA-106-B	3	40	49	2.0	2	3
SP-10 AIG Piping	SA-106-B	3	40	56	2.0	2	4
SP-11 AIG Piping	SA-106-B	3	40	62	3.0	2	4
SP-12 AIG Piping	SA-106-B	3	40	68	4.0	2	4
SP-13 AIG Piping	SA-106-B	3	40	75	4.0	2	5
SP-14 AIG Piping	SA-106-B	3	40	81	4.0	2	5
SP-15 AIG Piping	SA-106-B	3	40	33	2.0	2	2
SP-16 AIG Piping	SA-106-B	3	40	27	2.0	2	2
SP-17 AIG Piping	SA-106-B	3	40	20	2.0	2	1
SP-18 AIG Piping	SA-106-B	3	40	15	1.0	2	1
SP-19 AIG Piping	SA-106-B	3	40	36	2.0	2	2
SP-20 AIG Piping	SA-106-B	3	40	42	2.0	2	3
SP-21 AIG Piping	SA-106-B	3	40	48	2.0	2	3
SP-22 AIG Piping	SA-106-B	3	40	55	2.0	2	3
SP-23 AIG Piping	SA-106-B	3	40	61	2.0	2	4
SP-24 AIG Piping	SA-106-B	3	40	67	3.0	2	4
SP-25 AIG Piping	SA-106-B	3	40	74	3.0	2	4
SP-26 AIG Piping	SA-106-B	3	40	80	3.0	2	5
SP-27 AIG Piping	SA-106-B	3	40	87	4.0	2	5
SP-28 AIG Piping	SA-106-B	10	40	39	2.0	1	2

2.7 Combined cycle power plant STG vendor piping

2.7.1 STG vendor piping sheet 1

Line no.	Material	Size	Sch/ Thk	Pipe	SW BW	Valve	Boltup	Instrument	PS
Spray water system									
Pipe	SA-106-B	2	80	60	20	1	5		2
Gland Steam System (15Mo3)									
Pipe	15Mo3	6	80	140	30		1		15
Pipe	15Mo3	2.5	80	80	16		2		2
Pipe	15Mo3	5	80	40	8		1		1
Pipe	15Mo3	4	80	40	11		2	2	1
Pipe	15Mo3	6	80	60	12		2		2
Pipe	15Mo3	2	80	60	10		2		
Pipe	15Mo3	4	80	40	12		1		1
Pipe	15Mo3	6	80	140	30		1		21
Pipe	15Mo3	6	80	40	14	2	3	1	2
Pipe	15Mo3	2	80	140	42	1	1		24
Pipe	15Mo3	2	80	20	2		1		
Pipe	15Mo3	2	80	40	9		3		3
Leak-Off Steam System (15Mo3)									
Pipe	15Mo3	2	80	40	15		1		1
Pipe	15Mo3	2	80	40	15		1		
Pipe	15Mo3	3	80	100	15				16
Pipe	15Mo3	3	80	40	10		1		3
Pipe	15Mo3	4	80	40	8		1		2
Pipe	15Mo3	6	80	40	5				7
Pipe	15Mo3	6	80	20	8		1		
Lube oil supply									
Pipe-304L SS	304L SS	6	S10S	20	4		1		
Pipe-304L SS	304L SS	4	S10S	20	7		1		1
Pipe-304L SS	304L SS	2	S40S	40	13		1		1
Pipe-304L SS	304L SS	1	S40S	20	17		2		2

(Continued)

(Continued)

Line no.	Material	Size	Sch/ Thk	Pipe	SW BW	Valve	Boltup	Instrument	PS
Pipe-304L SS	304L SS	4	S10S	80	21		1		4
Pipe-304L SS	304L SS	2	S40S	20	9		1		
Pipe-304L SS	304L SS	2	S40S	20	9		1		
Pipe-304L SS	304L SS	4	S10S	40	13	1	1		2
Pipe-304L SS	304L SS	2	S40S	40	13		1		4

2.7.2 STG vendor piping sheet 2

Line no.	Material	Size	Sch/Thk	Pipe	SW BW	Valve	Boltup	Instrument	PS
Lube Oil Return									
Pipe-304L SS	304L SS	12	S10S	60	9	1	2		2
Pipe-304L SS	304L SS	6	S10S	20	4				
Pipe-304L SS	304L SS	5	S10S	20	3				
Pipe-304L SS	304L SS	6	S10S	40	8		1		2
Pipe-304L SS	304L SS	5	S10S	20	9		1		
Pipe-304L SS	304L SS	5	S10S	20	10		1		1
Pipe-304L SS	304L SS	10	S10S	40	8	1	1		2
Pipe-304L SS	304L SS	5	S10S	20	2				
Pipe-304L SS	304L SS	4	S10S	20	2				
Pipe-304L SS	304L SS	4	S10S	20	9		1		1
Pipe-304L SS	304L SS	2.5	S10S	20	9	1	1		1
Hydrostatic oil, oil purification									
Pipe-304L SS	304L SS	1.25	S40S	40	20				4
Pipe-304L SS	304L SS	2	S40S	40	15	1			2
Pipe-304L SS	304L SS	2	S40S	40	4				5

(*Continued*)

(Continued)

				SW					
Line no.	Material	Size	Sch/Thk	Pipe	BW	Valve	Boltup	Instrument	PS
Pipe-304L SS	304L SS	2.5	S10S	80	22		2		17
Jacking oil									
Pipe-304L SS	304L SS	20	S10S	60	15	3	2		27
Pipe-304L SS	304L SS	12	S10S	120	8				11
Pipe-304L SS	304L SS	2.5	S10S	120	32				24
Oil flushing									
Rubberized hose	304L SS	0.75	S40S	20			26		
External hydraulic oil									
Pipe-SS	304L SS	3	S10S	220	60				60
Pipe-SS	304L SS	2	S40S	40	20				15
Pipe-SS	304L SS	2	S40S	240	80				60
Pipe-SS	304L SS	1.5	S40S	200	60				50

2.7.3 STG vendor piping sheet 3

			SW					
Line no.	Material Size	Sch/Thk	Pipe	BW	Valve	Boltup	Instrument	PS
Condensate	3	std	60	11	2	5		1
Condensate	2	40 s	20	8				1
Condensate	1.5	80	40	8		1		1
Condensate	1.5	80	40	8		1		1
Condensate	1.5	80	40	8		1		1
Steam turbine drain	3	80	60	5		2		1
Steam drain	6	std	40	2		1		1
Low-pressure steam	1.5	80	40	8		1		1
Steam turbine Drain	6	80	40	6		2		1
Steam drain	6	std	40	6		2		1
Steam drain	1	80	40	6		2		2
Steam drain	1	80	40	6		2		2
Steam drain	1	80	40	6		2		2
Steam drain	1	80	40	6		2		2
Steam drain	1	80	40	6		2		2
Condensate	3	std	80	16	2	10		2
Boiler feedwater	3	std	60	12	2	4		1
Boiler feedwater	3	std	60	12	2	4		1
Steam drains	6	std	60	10.2	4	10		2
Steam drains	1	80	60	10.2		2		2
Steam drains	1	80	60	10.2		2		2
Steam drains	1	80	60	10.2		2		2
Steam drains	1	80	60	10.2		2		2
Steam drains	1.5	80	60	10.2		2		2
Steam drains	1.5	80	60	10.2		2		2
Steam drains	1.5	80	60	10.2		2		2

(*Continued*)

(Continued)

Line no.	Material Size	Sch/Thk	Pipe	BW	SW Valve	Boltup	Instrument	PS
Steam drains	1.5	80	60	14		2		2
Trim	1	80	20	10	2	2		1
Condensate	1	80	60	2	2	2		2
Steam drains	24	std	20	8		2		1
Low-pressure steam	1	80	40	8		2		2
Low-pressure steam	1	80	40	8		2		2
Low-pressure steam	1.5	80	40	8		2		2
Low-pressure steam	1.5	80	40	8		2		2
Low-pressure steam	2	80	40	8		2		2
Low-pressure steam	1.5	80	40	16		2		2
Low-pressure steam	2	80	40	8		2		2
Steam drains	2	80	40	8		2		2
Steam drains	2	80	40	8		2		2
Steam drains	2	80	40	8		2		2
Steam drains	2	80	40	14		2		2
Steam turbine drain	2	80	40	8		2		2
Steam drains	2	80	40	8		2		2
Steam drains	2	80	40	8		2		2
Hot reheat steam	2	80	40	8		2		2
Hot reheat steam	2	80	40			2		2

Chapter 3

Construction database system

3.1 Introduction

Estimating data systems uses historical data from quantity takeoff quantities and man-hours for estimating new projects of similar equipment or systems. Equipment in process plants, regardless of manufacturer, has processes similar to previously designed units where only the size and weight change. The duplication of design enables the estimator to set up cost codes, summarize and analyze historical data and trends, and maintain craft man-hour databases. This chapter will enable the estimator to collect, verify, and analyze historical data to develop man-hour tables that can be used to set up and implement a man-hour database system using comparable cost and man-hour data to estimate future projects.

3.2 Construction database

Collection of historical data must be a systematic method of categorizing the information using a standard structure referred to as a code of accounts. Before you can effectively collect and analyze historical data, you must have a coding system in place. The job cost by cost code and type is used to describe the task and cost type to describe the man-hours associated with the task. After historical data has been collected, it must be verified by graphic and statistical analysis. Evaluation of the data reduces inconsistencies within the data and correlation will determine if the data is acceptable. This step is critical in the development of a database. The field provides the field data report and the estimator develops a man-hour database that is used to estimate and validate historical data to estimate future projects.

3.3 Development of industrial construction database

"Construction-based estimating data" is an estimating database that uses historical data from quantity takeoff and direct craft labor man-hours as a tool for bidding new projects of similarly designed equipment or systems. Most equipment and systems in industrial process plants have designs similar to

Industrial Construction Estimating Manual. DOI: https://doi.org/10.1016/B978-0-12-823362-7.00003-X
43

previously designed systems except for size and weight. This duplication of design enables the estimator to cost code, collect, summarize, and verify field data to develop a man-hour database to estimate new projects of similar design.

Development of construction-based estimating database for industrial process plants:

- The quantity takeoff is developed during the bid preparation and quantifies the materials.
- The scope of work is defined and the erection sequence is based on the duplication of design.
- Cost codes are used to describe the task and cost type to describe task man-hours.
- Two methods for the measurement of construction time are used to compile the man-hour data
  - nonrepetitive one-cycle time study
  - field data report.
- The detailed estimate is prepared by combining the unit method and unit quantity method
- unit method
- unit quantity model.
- Excel estimate spreadsheet, based on the Unit Quantity Model, calculates the craft labor.

3.3.1   Benefits of unit man-hour database

The database creates efficiency, quality, and speed and with software is accessible to everyone.

- It uses historical data from field civil, structural, piping, and mechanical work.
- It develops historical data from material takeoff quantities.
- It keeps accurate records of man-hour labor productivity and revises the database constantly.
- It uses labor man-hours as a tool for estimating new projects of similar mechanical and civil systems.
- It compares data similarities to actual cost and time spent on work activities with those of the estimate.
- It provides an accurate basis for determining labor productivity.
- It tracks field-labor productivity for civil, structural, piping, and mechanical to build a historical database.
- It tracks the cost basis that provides transparency for change orders eliminating disputes.

● It publishes company man-hour tables that must be validated and clarified using statistical analysis.

3.4 Piping man-hour database

The development of underground cast iron piping database for industrial process plants:

The historical data is collected on site every day and summarized in a spreadsheet to be verified by regression analysis.

Based on data collected in the field on Project X, the unit rate summary for pipe set and align CI Pipe is summarized in the next.

3.4.1 Pipe set and align, cast iron—lead and mechanical joint

Facility—industrial plant (underground drainage piping)

Pipe size (in.)	Pipe set and align (MH/lf)
4	0.12
6	0.18
8	0.24
10	0.30
12	0.36
16	0.48
18	0.54
20	0.60
24	0.72

Pipe man-hours include handle, haul, place, and align in trench

3.4.2 Verify historical data for cast iron pipe—pipe set and align

Facility—industrial plant (underground drainage piping)
Data for input: Pipe set and align, cast iron—lead and mechanical joint
Man-hour (Y): R1 = 0.12, 0.18, 0.24, 0.30, 0.36, 0.48, 0.54, 0.60, 0.72
Pipe size, inches (X): R2 = 4, 6, 8, 10, 12, 16, 18, 20, 24
Man-hour per lf (Fig. 3.1)

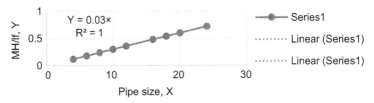

FIGURE 3.1 Pipe set and align, cast iron.

X	Y
Pipe size	MH/lf
4	0.12
6	0.18
8	0.24
10	0.30
12	0.36
16	0.48
18	0.54
20	0.60
24	0.72

Covar (R1, R2)	1.23
Varp (R2)	0.04
Slope (R1, R2)	0.0300
Intercept (R1, R2)	0.0000

The coefficient of determination, R^2, is exactly $+1$ and indicates a positive fit. All data points lie exactly on the straight line. The relationship between X and Y variables is such that as X increases, Y also increases.

3.4.3 Mechanical joint MH/JT, cast iron—lead and mechanical joint

Facility—industrial plant (underground drainage piping)

Pipe size (in.)	Mechanical joint (MH/JT)
4	0.80
6	1.20
8	1.60
10	2.00
12	2.40
16	3.20
18	3.60
20	4.00
24	4.80

Pipe man-hours include handle, haul, place, and align in trench

3.4.4   Verify historical data for cast iron pipe—mechanical joint MH/JT

Facility—industrial plant (underground drainage piping)

   Data for Input: Mechanical joint MH/JT, cast iron—lead and mechanical joint

   Man-hour (Y): R1 = 0.80, 1.20, 1.60, 2.00, 2.40, 3.20, 3.60, 4.00, 4.80
   Pipe size, inches (X): R2 = 4, 6, 8, 10, 12, 16, 18, 20, 24
   Man-hour per lf (Fig. 3.2)

X	Y
Pipe size	MH/JT
4	0.80
6	1.20
8	1.60
10	2.00
12	2.40
16	3.20
18	3.60
20	4.00
24	4.80

Covar (R1, R2)	8.20
Varp (R2)	1.64
Slope (R1, R2)	0.2000
Intercept (R1, R2)	0.0000

The coefficient of determination, R^2, is exactly $+1$ and indicates a positive fit. All data points lie exactly on the straight line. The relationship between X and Y variables is such that as X increases, Y also increases.

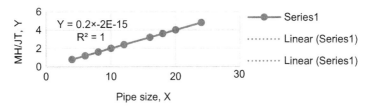

FIGURE 3.2   Mechanical joint, cast iron.

3.4.5 Handle and install fitting, cast iron—lead and mechanical joint

Facility—industrial plant (underground drainage piping)

Pipe size (in.)	Handle install fitting (MH/ea.)
4	0.24
6	0.36
8	0.48
10	0.60
12	0.72
16	0.96
18	1.08
20	1.20
24	1.44

Pipe man-hours include handle, haul, place, and align in trench

3.4.6 Verify historical data for cast iron pipe—handle and install fitting

Facility—industrial plant (underground drainage piping)

Data for input: Handle and install fitting, cast iron—lead and mechanical joint

Man-hour (Y): R1 = 0.24, 0.36, 0.48, 0.60, 0.72, 0.96, 1.08, 1.20, 1.44
Pipe size, inches (X): R2 = 4, 6, 8, 10, 12, 16, 18, 20, 24
Man-hour per lf (Fig. 3.3)

X	Y
Pipe size	MH/ea.
4	0.24
6	0.36
8	0.48
10	0.60
12	0.72

(Continued)

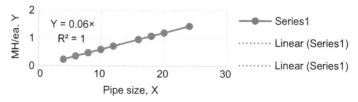

FIGURE 3.3 Handle and install fitting, cast iron.

(Continued)

X	Y
Pipe size	MH/ea.
16	0.96
18	1.08
20	1.20
24	1.44

Covar (R1, R2)	2.46
Varp (R2)	0.15
Slope (R1, R2)	0.0600
Intercept (R1, R2)	0.0000

The coefficient of determination, R^2, is exactly $+1$ and indicates a positive fit. All data points lie exactly on the straight line. The relationship between X and Y variables is such that as X increases, Y also increases.

3.4.7   Verification of cast iron underground piping by regression analysis

- Pipe set and align, MH/lf; $R^2 = 1$
- Mechanical joint, MH/JT; $R^2 = 1$
- Handle and install fitting, MH/ea.; $R^2 = 1$

$R^2 = 1$ indicates the model and fits the data.

The historical data collected, summarized, and analyzed using the regression model is verified and the estimator sets up the unit man-hour table, for cast iron piping, to include in the piping database.

3.4.8   Handle and install pipe, cast iron—lead and mechanical joint

Facility—industrial plant (underground drainage piping)

Pipe size (in.)	Pipe set and align (MH/lf)	Lead and mechanical joint (MH/JT)	Handle install fitting (MH/ea.)
4	0.12	0.80	0.24
6	0.18	1.20	0.36
8	0.24	1.60	0.48
10	0.30	2.00	0.60
12	0.36	2.40	0.72
16	0.48	3.20	0.96
18	0.54	3.60	1.08

(Continued)

(Continued)

Pipe size (in.)	Pipe set and align (MH/lf)	Lead and mechanical joint (MH/JT)	Handle install fitting (MH/ea.)
20	0.60	4.00	1.20
24	0.72	4.80	1.44

Pipe man-hours include handle, haul, place, and align in trench

3.5 Illustrative examples of database for civil, structural and miscellaneous steel, and pipeline

3.5.1 Civil database for hydrogen plant

Facility: Hydrogen plant—earthwork		
Earthwork	MH	Unit
Mass machine excavation—depth to 6′	0.339	cy
Machine excavation for single foundations	0.400	cy
Mass machine backfill	0.470	cy
Machine backfill foundations	0.737	cy
Dispose of surplus material	0.355	cy
Finished grading with 6″ thick gravel	0.050	sf
Access roads heavy substructure and 6″ thick gravel	0.084	sf
Ground water lowering	0.028	sf
Facility: Hydrogen plant—concrete works		
Concrete works	MH	Unit
Blinding concrete	0.013	sf
Foundation slabs—4000PSI	1.831	cy
Footing concrete—4000PSI	2.310	cy
Pier concrete—4000PSI	2.751	cy
Elevated floor slab concrete	3.850	cy
Concrete walls Thk. 8″–10″	3.852	cy
Reinforcement bars	0.007	lb
PVC membrane, 0.2 mm	0.002	sf
Formwork	0.187	sf
Extra over for formwork	0.088	sf
Anchor bolts 6/8 to 2″ hot-dip galv.	0.088	lb
Nonshrink grout (cementitious)	0.242	CF
Epoxy grout, high strength	0.827	CF
Exp joints	0.109	lf
Water stop	0.110	lf
Embedded items	0.088	lb
Extra over for compressive strength	0.088	cy
Templates for columns and vessels	0.088	lb
Compressor concrete	1.955	cy
Compressor Rebar	13.800	ton

Facility: Hydrogen plant—miscellaneous

Miscellaneous	MH	Unit
PVC cable conduits d = 6″, underground	1.1	lf
Sewer system, PVC pipe d = 6″/8″	1.1	lf
Fencing 8′ high with five lines of barbed wire and one 20′ wide gate	0.198	lf
Gates for access roads	25.5	ea.
Fencing for transformer bays, including two doors, framed wire mesh	0.715	lf
Oil-resistant paint	0.053	sf
Storm water U/G RCP 12″ diameter	1	lf
Storm water U/G RCP 18″ diameter	1.5	lf
Manholes, cast-in-place	8	ea.
Catch basins, cast-in-place	12	ea.
Outlet structure, cast-in-place	240	ea.

3.5.2 Facility—industrial plant (simple foundation)

	MH/ cy	MH/ sf	MH/ lf	MH/ lb	MH/diameter in. ft.
Structure excavation—backhoe	0.20				
Structure excavation—hand	2.80				
Structure backfill and compact—loader and Wacker	0.60				
Structure backfill and compact—hand	3.00				
Edge forms—slabs and foundations			0.10		
Fabricate, install, strip foundation forms—1 use		0.30			
Fabricate, install, strip pedestal forms—1 use		0.25			
Fabricate, install, strip wall forms—1 use		0.20			
Fabricate and install reinforcing steel				0.03	
Layout templates and set anchor bolts					0.25
Set embedded steel—curb angle, etc.				0.04	
Place concrete—from truck below grade	0.80				
Place concrete—slabs at grade	1.00				
Place concrete—pedestals and walls	2.00				
Finish flat concrete		0.12			
Patch and sack concrete		0.05			
Install mesh			0.02		

3.5.3 Structural and miscellaneous steel database for combined cycle power plant

Facility: Combined cycle power plant—structural steel		
HRSG utility steel bridge	**MH**	**Unit**
Light—0–19 lb/ft.; ≤20 ton	28.0	ton
Medium—20–39 lb/ft.; ≥20 ton	18.0	ton
Heavy—40–79 lb/ft.; >100 ton	13.2	ton
X heavy—80–120 lb/ft.; ≥20 ton	12.0	ton
STG utility bridge steel		
Light—0–19 lb/ft.; ≤20 ton	28.0	ton
Medium—20–39 lb/ft.; ≥20 ton	18.0	ton
Medium—20–39 lb/ft.; ≥20 ton	18.0	ton
X heavy—80–120 lb/ft.; ≥20 ton	12.0	ton
STG utility bridge steel inside enclosure		
Light—0–19 lb/ft.; ≤20 ton	28.0	ton
Medium—20–39 lb/ft.; ≤20 ton	18.0	ton
Heavy—40–79 lb/ft.; > 100 ton	13.2	ton
X heavy—80–120 lb/ft.; ≤20 ton	12.0	ton
GSU transformer access platforms		
Light and medium	26.0	ton
Stair tower and ladders		
Light and medium	26.0	ton
Stair treads	0.85	ea.
Caged ladders	0.35	lf
Iso phase support steel		
Light and medium	26.0	ton
Dead end structure		
Light and medium	26.0	ton
230 kV switchyard structure including bus supports		
Light and medium	26.0	ton
Transmission line poles		
Light and medium	26.0	ton
Grating, handrail, and toe plate		
2″ serrated floor grating	0.20	sf
1.5″ serrated floor grating	0.20	sf
Handrail and toe plate	0.25	lf

3.5.4 Pipeline database

Underground piping labor units	Labor	Minimum
C.S.B. weld to 0.500 Thk.	0.35 MH/diameter in.	1.00 MH/weld
Handling	0.012 MH/diameter in. ft.	0.08 MH/lf
Tape wrap welds	0.10 MH/diameter in.	0.50 MH/JT
Boltups	0.15 MH/bolt	0.60 MH/boltup
Valves and specialties	0.20 MH/diameter in.	0.20 MH/valve

(Continued)

(Continued)

Underground piping labor units	Labor	Minimum
Above ground piping labor units		
C.S.B. weld to 0.500 Thk.	0.40 MH/diameter in.	1.00 MH/weld
Handling	0.02 MH/diameter in. ft.	0.08 MH/lf
Boltups	0.20 MH/bolt	0.80 MH/boltup
Valves and specialties 150# and 300#	0.20 MH/diameter in.	0.20 MH/valve
Valves and specialties 600# and 900#	0.40 MH/diameter in.	0.40 MH/valve
O-lets and nozzles	2.0 × butt weld	
Tie-Ins	3.0 × butt weld	
Socket welds	0.50 MH/diameter in.	0.80 MH each
Screwed joints	0.20 MH/diameter in.	0.20 MH/JT
Trench work for piping	Labor man-hours	
Excavation:		
Large trenches (75 + cy per location)	0.04 MH/cy	
Intermediate trenches (20–74 cy per location)	0.25 MH/cy	
Small trenches (less than 20 cy per location)	0.50 MH/cy	
Backfill and compaction:		
Large volume (75 + cy per location)	0.04 MH/cy	
Intermediate volume (20–74 cy per location)	0.40 MH/cy	
Small trenches (less than 20 cy per location)	0.67 MH/cy	
Load, haul, and dump:		
Dump site within 30 min of job	0.04 MH/cy	
Sand bedding and cover	0.25 MH/cy	

3.5.5 Schedule A—combined cycle power plant piping

Standard labor estimating units		
Facility—combined cycle power plant	Large bore piping	Small bore piping
	Unit of measure	Unit of measure
Description	Man-hours per unit	Man-hours per unit
Handle and install pipe, carbon steel, welded joint	Diameter inch feet	MH/lf
WT ≤ 0.375"	0.07	0.18
0.406" ≤ WT ≤ 0.500"	0.09	0.23
0.562" ≤ WT ≤ 0.688"	0.11	0.28
0.718" ≤ WT ≤ 0.938"	0.14	0.35
1.031" ≤ WT ≤ 1.219"	0.20	0.50

(*Continued*)

(Continued)

Standard labor estimating units

Facility—combined cycle power plant	Large bore piping	Small bore piping
	Unit of measure	Unit of measure
Description	Man-hours per unit	Man-hours per unit
1.250" ≤ WT ≤ 1.312"	0.25	0.75
Welding butt welds, carbon steel, Arc-uphill	Diameter inch	MH/ea.
WT ≤ 0.375"	0.50	1.10
0.406" ≥ WT ≤ 0.500"	0.55	1.20
0.562" ≥ WT ≤ 0.688"	1.05	2.20
0.718" ≥ WT ≤ 0.938" (PWHT)	1.20	2.45
1.031" ≥ WT ≤ 1.219" (PWHT)	1.45	2.70
1.250" ≤ WT ≤ 1.312" (PWHT)	2.20	4.40
Olet—SOL, TOL, WOL	2 × BW	2 × BW
Stub in	1.5 × BW	1.5 × BW
Socketweld		Per SW table
PWHT craft support labor	0.45	1.00
Boltup of flanged joints by weight class	Diameter inch	MH/ea.
150#/300# boltup	0.40	1.00
600#/900# boltup	0.50	1.20
1500#/2500# boltup	0.65	1.60
Handle valves by weight class	Diameter inch	MH/ea.
150# and 300# manual valves	0.45	1.00
600# and 900# manual valves	0.90	1.80
Heavier manual valve ≥ 1500#	1.80	2.00

3.6 Balance of plant equipment estimating database

3.6.1 Set, align, couple, and grout pumps

Description	MH	Unit
Pumps—set, align, couple, and grout		
Motor horsepower less than 15	20.00	ea.
Motor horsepower 15–30	2.10	HP
Motor horsepower 31–50	1.80	HP
Motor horsepower 51–100	1.60	HP
Motor horsepower 101–125	1.40	HP
Motor horsepower 126–300	1.30	HP
Motor horsepower 301–500	1.00	HP
Motor horsepower more than 500	0.80	HP
Vacuum pumps—motor horsepower 0–30	1.00	HP

(Continued)

(Continued)

Description	MH	Unit
Table-Rig, set, and align, weld tanks and vessels		
Horizontal tanks and vessels		
Weight range tons 0−5	60	ea.
Weight range tons 6−10	90	ea.
Weight range tons 11−20	140	ea.
Table-Rig, set, and align factory assembled skids		
Skid-mounted unit weight (lb) 0−3000	28	ea.
Skid-mounted unit weight (lb) 3001−6000	40	ea.
Skid-mounted unit weight (lb) 6001−15,000	60	ea.
Table-Rig, set, and align heater		
Preheater and heater weight (lb) 0−500	30	ea.
Preheater and heater weight (lb) 501−1500	50	ea.
Preheater and heater weight (lb) 1501−2000	80	ea.
Table-Rig, set, and align compressors		
Fuel gas compressor 500 HP, weight 45,000 lb	27.6	ton
Gas compressor cooler-package unit 1 and 2 0−2000 lb	60	ea.
Air compressor A/B 0−5000 lb	100	ea.
Table-Rig, set, and align heat exchangers		
Closed cooling water heat exchanger A/B/C	60	ea.
Tale-Rig, set, and align eyewash/shower		
Eyewash/shower	32	ea.

Chapter 4

Construction labor estimate

4.1 Introduction

This chapter presents the principles of the labor estimate. Labor, material, and equipment cost are the basis for the detailed labor estimate, and these elements must be determined efficiently. Detailed labor estimates determine the direct craft labor and man power requirements for the project. The initial labor estimate is the basis for the estimator to obtain the project schedules. Estimators use statistical and mathematical equations, and construction databases, which are entered into the computer, that model the work of construction. The estimator must set up cost codes, summarize, and analyze labor productivity and trends, and compare data similarities to actual cost and time spent on work activities with those of the estimate.

Construction-work estimating data sets, or books, for construction analysis and cost estimating are popular sources of information. With all the combinations of construction and engineering design, labor skill, material, construction equipment, and methods and procedures, the contractor should develop and maintain their own data sets. Estimating data must be verified by statistical analysis, field data reports and the data must be updated and maintained.

Construction-work estimating references are useful as a secondary resource if the data is verified by analysis and the work scope is comparable to future content.

4.2 Elements of construction-work estimate

- The *quantity takeoff*: It is developed during the bid preparation and quantifies the materials required to complete the project.
- Labor *hour*: This basic unit is defined as one worker working for 1 hour.
- Labor *cost*: Man-hour × labor rate.
- Material *cost*: It is to ensure a complete cost of material requirements associated with the installation of the materials.
- Equipment *cost*: Cost for rental rates for third party equipment and company-owned equipment.
- *Subcontractor quotes*: Cost for subcontractors to complete parts of the construction.

Industrial Construction Estimating Manual. DOI: https://doi.org/10.1016/B978-0-12-823362-7.00004-1

4.3 Construction-work estimating

The work estimate is created using construction-work elements identified in the project material estimate. Estimating worksheets are used to develop the labor estimate.

Unit-quantity method—model for estimating
The model for detailed estimating based on the unit-quantity method:

The unit-quantity model is given by: $MH\mu = \sum n_i(MH_a)$

where $MH\mu$ is the quantity n_i, unit man-hours for field construction work; n_i is the task takeoff quantity I, in dimensional units; MH_a is the average unit man-hour per quantity n_i; μ is the Task 1, 2, k from quantity takeoff associated w/construction work.

The n_i quantity is the takeoff for the field scope of work.

MH_a man-hours are determined from field construction man-hour tables.

The model focuses on the quantity n_i. MH_a is the required unit man-hour to install takeoff field task determined from field construction labor tables. Many different work variations can be illustrated with the unit-quantity method.

4.4 Four scopes of work illustrate estimate worksheets using the unit-quantity model

4.4.1 Combined cycle power plant; boiler foundations—HRSG Stack and HRSG

Work; material takeoff	Size			Quantity		Unit MH	Total MH
	L	W	H	n_i	Unit	MH_a	$n_i MH_a$
Excavation	200	50	5	2037.0	cy	0.400	814.7
Forms	204	55	5	2580.0	sf	0.350	903.0
Rebar				107.0	ton	14.0	1498.0
Concrete	200	50	5	1944.4	cy	2.0	3888.9
Mud mat	204	54	0.33	136.0	cy	0.500	68.0
Backfill, compact				80.0	cy	0.631	50.5
Column total							**7223.1**

4.4.2 Combined cycle power plant—structural and miscellaneous steel

Description	Unit MH n_i	Unit	MH_a	Total MH n_i MH_a
HRSG utility steel bridge	**204.6**	**ton**		**3803.8**
Light—0–19 lb/ft.	11.0	ton	24	264
Medium—20–39 lb/ft.	33.0	ton	23	759
Heavy—40–79 lb/ft.	105.6	ton	18	1900.8
X heavy—80–120 lb/ft.	55.0	ton	16	880
STG utility bridge steel	**187.0**	**ton**		**3452.5**
Light—0–19 lb/ft.	8.4	ton	24	201.6
Medium—20–39 lb/ft.	27.9	ton	23	641.7
Heavy—40–79 lb/ft.	99.0	ton	18	1782
X heavy—80–120 lb/ft.	51.7	ton	16	827.2
STG utility bridge steel inside enclosure	**37.4**	**ton**		**690.8**
Light—0–19 lb/ft.	1.7	ton	24	40.8
Medium—20–39 lb/ft.	5.6	ton	23	128.8
Heavy—40–79 lb/ft.	19.8	ton	18	356.4
X heavy—80–120 lb/ft.	10.3	ton	16	164.8
GSU transformer access platforms	**13.5**	**ton**		**310.5**
Light and medium	13.5	ton	23	310.5
Stair tower and ladders	**14.2**	**ton**		**464.2**
Light and medium	10.4	ton	23	239.2
Stair treads	300.0	ea.	0.55	165
Ladders	200.0	lf	0.30	60
Iso-phase support steel	**33.0**	**ton**		**759**
Light and medium	33.0	ton	23	759
Dead-end structure	**33.0**	**ton**		**759**
Light and medium	33.0	ton	23	759
230 kV Switchyard structure including bus supports	**16.5**	**ton**		**379.5**
Light and medium	16.5	ton	23	379.5
Transmission line poles	**6.0**	**ton**		**138**
Light and medium	6.0	ton	23	138
Grating, handrail, and toe plate	**17.1**	**ton**		**1080**
2″ serrated floor grating	11,200.0	sf	0.05	560
1.5″ serrated floor grating	4100.0	sf	0.05	205
Handrail and toe plate	2100.0	sf	0.15	315
Total structural and miscellaneous steel	**562.2**	**ton**		**11,837.3**
MH/ton	**21.1**			

GSU, Generator step-up transformer; HRSG, Heat recovery steam generator; STG, steam turbine generator.

4.4.3 Refinery—vessels/columns

Description	Quantity		Unit MH	Total MH
	n_i	Unit	MH_a	n_i MH_a
Solvent absorber 8′ − 6″ D × 69′ − 0″, 319.7 ton	**319.7**	**ton**		**1839**
Unload, handle, haul up to 2000′, rig, set and align, make up Fdn AB	319.7	ton	1.9	600
Install platforms	12	ea.	32	384
Install caged ladders	100	lf	0.35	35
Remove and replace manway cover (24″ 300# Removable-Davit)	1	ea.	40	40
Install double downflow valve trays (16 trays)	16	ea.	40	640
Install demisting pads (single grid-support, pad, grid-top)	1	ea.	60	60
Vortex breaker	1	ea.	32	32
Packing (pall rings)	1	lot	48	48
Rich solvent flash drum 7′ − 0″ D × 24′ − 0″; 14.9 ton	**14.9**	**ton**		**416**
Unload, handle, haul up to 2000′, rig, set and align, make up Fdn AB	14.9	ton	10.8	160
Install platforms and ladders	2	ea.	32	64
Remove and replace manway cover (24″ 300# Hinged)	1	ea.	40	40
Inlet box	1	ea.	60	60
Vortex breaker	1	ea.	32	32
Install demisting pads (single grid-support, pad, grid-top)	1	ea.	60	60
Solvent regenerator 8′ − 0″ D × 72′ − 6″; 39 ton	**39**	**ton**		**1140**
Unload, handle, haul up to 2000′, rig, set and align, make up Fdn AB	39.0	ton	8.2	320
Install platforms and ladders	5	ea.	32	160
Remove and replace manway cover (24″ 300# Removable-Davit)	1	ea.	40	40
Install double downflow valve trays (12 trays)	12	ea.	40	480
Install demisting pads (single grid-support, pad, grid-bottom)	1	ea.	60	60
Vortex breaker	1	ea.	32	32
Packing (pall rings)	1	lot	48	48
Solvent regenerator reflux drum 5′ − 0″ D × 15′ − 0″; 2.6 ton	**2.6**	**ton**		**276**
Unload, handle, haul up to 2000′, rig, set and align, make up Fdn AB	2.6	ton	30.6	80
Install platforms and ladders	2	ea.	32	64
Remove and replace manway cover (24″ 300# Hinged)	1	ea.	40	40
	1	ea.	60	60

(Continued)

(Continued)

Description	Quantity n_i	Unit	Unit MH MH_a	Total MH n_i MH_a
Install demisting pads (single grid-support, pad, grid-top)				
Vortex breaker	1	ea.	32	32
Total vessels/columns	**376.2**	**ton**		**3671**
MH/ton	9.8			

4.4.4   Gasifier—Feedstock Bunker A and B

Description	Quantity n_i	Unit	Unit MH MH_a	Total MH n_i MH_a
Feedstock Bunker A and B				**7449**
Feedstock Bunker A 40′ D × 101′	**231**	**ton**		**3724**
Receive, transport material, lift, and set	231	ton	1.4	320
Handle sheets per ring 6 ea. × 40′ × 8 − 3′ total 9 rings × 6 sheets = 54 sheets	54	ea.	12	648
Fit and weld vertical 0.75″ WT CS 6 × 8′ − 3″ × 9 rings 448 weld DI	448	DI	1.1	504
Fit and weld horizontal 0.75″ WT CS 8 × 251.3′ 2011 weld DI	2011	DI	0.35	704
Overlay stainless steel liner	1	job	1548	1548
Feedstock Bunker B 40′ D × 101′	**231**	**ton**		**3724**
Receive, transport material, lift, and set	231	ton	1.4	320
Handle sheets per ring 6 ea. × 40′ × 8 − 3′ total 9 rings × 6 sheets = 54 sheets	54	ea.	12	648
Fit and weld vertical 0.75″ WT CS 6 × 8′ − 3″ × 9 rings 448 weld DI	448	DI	1.1	504
Fit and weld horizontal 0.75″ WT CS 8 × 251.3′ 2011 weld DI	2011	DI	0.35	704
Overlay stainless steel liner	1	job	1548	1548

4.5   Typical process piping estimate

4.5.1   Field Erect-HRSG HP piping and supports

Facility—Combined cycle power plant
   Triple-pressure w/Reheat for F-Class GT-Three Wide

4.5.2   Estimate sheet—Handle and install pipe-welded joint

Description	Size	Pipe handle lf	DIF	MH/ DIF	PF MH	Hydro DIF	MH/ DIF	PF MH	Total MH
HP piping and supports									
Sch 140/ 0.906", SA-106 B, HP-03 Econ 1 to HP Econ 2	8	25	200	0.14	28	200	0.134	27	55
Sch 160/ 0.906", SA-106 B, HP-03 Econ 1 to HP Econ 2	6	42	252	0.14	35	252	0.101	25	61
Sch 140/ 0.812", SA-106 B, HP-04 Econ 2 to HP steam drum	8	96	768	0.14	108	768	0.134	103	211
Sch 160/ 0.906", SA-106 C, HP-04 Econ 2 to HP steam drum	6	1	6	0.14	1	6	0.101	1	1
Sch 160/ 0.906", SA-106 B, HP-09 steam drum to HP SH 1	6	24	144	0.14	20	144	0.101	15	35
Sch 160/ 0.906", SA-106 B, HP-10 steam drum to HP SH 1	6	23	138	0.14	19	138	0.101	14	33
Sch 160/ 0.906", SA-106 B, HP-11 steam drum to HP SH 1	6	23	138	0.14	19	138	0.101	14	33
	6	23	138	0.14	19	138	0.101	14	33

(*Continued*)

(Continued)

Description	Size	Pipe handle lf	DIF	MH/ DIF	PF MH	Hydro DIF	MH/ DIF	PF MH	Total MH
Sch 160/ 0.906″, SA-106 B, HP-12 steam drum to HP SH 1									
Sch 160/ 0.906″, SA-106 B, HP-13 steam drum to HP SH 1	6	23	138	0.14	19	138	0.101	14	33
Sch 160/ 0.906″, SA-106 B, HP-14 steam drum to HP SH 1	6	23	138	0.14	19	138	0.101	14	33
1.25″ WT, SA-335 P91, HP-15 SHTR 1 to HP SHTR 2	10	68	680	0.25	170	680	0.240	163	333
1.031″ WT, SA-335 P91, HP-15 SHTR 1 to HP SHTR 2	8	3	24	0.20	5	24	0.192	5	9
SA-335 P91, HP-15 SHTR 1 to HP SHTR 2	3	1	3	0.20	1	3		0	1
1.25″ WT, SA-335 P91, HP-16 SHTR 1 to HP SHTR2	10	68	680	0.25	170	680	0.240	163	333
1.032″ WT, SA-335 P91, HP-16 SHTR 1 to HP SHTR 2	8	3	24	0.20	5	24	0.192	5	9
	3	1	3	0.20	1	3		0	1

(Continued)

(Continued)

Description	Size	Pipe handle			PF	Hydro		PF	Total
		If	DIF	MH/DIF	MH	DIF	MH/DIF	MH	MH
SA-335 P91, HP-16 SHTR 1 to HP SHTR 2 1.25" WT, SA-335 P91, HP-17 SHTR 1 to HP SHTR 2	10	68	680	0.25	170	680	0.240	163	333
1.031" WT, SA-335 P91, HP-17 SHTR 1 to HP SHTR 2	8	3	24	0.20	5	24	0.192	5	9
SA-335 P91, HP-17 SHTR 1 to HP SHTR 2	3	1	3	0.20	1	3		0	1
Column total		519	4181		815	4181		744	1558

Facility—Combined cycle power plant
Triple-pressure w/Reheat for F-Class GT-Three Wide

4.5.3 Estimate sheet 2—Welding: BW, SW, PWHT arc-uphill

Description	Size	BW	SW		BW	SW		PWHT	PF
		JT	JT	DI	MH/DI	MH/JT	Factor	MH/JT	MH
HP piping and supports									
Sch 140/0.906", SA-106 B, HP-03 Econ 1 to HP Econ 2	8			0	1.20	1.4	1.0	0	0
Sch 160/0.906", SA-106 B, HP-03 Econ 1 to HP Econ 2	6	9	9	54	1.20	1.4	1.0	0.5	102

(Continued)

(Continued)

Description	Size	BW JT	SW JT	DI	BW MH/ DI	SW MH/ JT	Factor	PWHT MH/ JT	PF MH
Sch 140/ 0.812", SA-106 B, HP-04 Econ 2 to HP steam drum	8	4	6	32	1.20	1.4	1.0	0.5	61
Sch 160/ 0.906", SA-106 C, HP-04 Econ 2 to HP steam drum	6	3		18	1.20	1.4	1.0	0.5	30
Sch 160/ 0.906", SA-106 B, HP-09 steam drum to HP SH 1	6	4	4	24	1.20	1.4	1.0	0.5	45
Sch 160/ 0.906", SA-106 B, HP-10 steam drum to HP SH 1	6	3	4	18	1.20	1.4	1.0	0.5	35
Sch 160/ 0.906", SA-106 B, HP-11 steam drum to HP SH 1	6	4	4	24	1.20	1.4	1.0	0.5	45
Sch 160/ 0.906", SA-106 B, HP-12 steam drum to HP SH 1	6	3	4	18	1.20	1.4	1.0	0.5	35
Sch 160/ 0.906", SA-106 B, HP-13 steam drum to HP SH 1	6	4	4	24	1.20	1.4	1.0	0.5	45
Sch 160/ 0.906", SA-106 B, HP-14 steam drum to HP SH 1	6	3	4	18	1.20	1.4	1.0	0.5	35
1.25" WT, SA-335 P91, HP-15 SHTR 1 to HP SHTR 2	10	4	5	40	2.2	1.4	2.8	0.5	279

(Continued)

(Continued)

Description	Size	BW JT	SW JT	DI	BW MH/DI	SW MH/JT	Factor	PWHT MH/JT	PF MH
1.031″ WT, SA-335 P91, HP-15 SHTR 1 to HP SHTR 2	8	2		16	1.45	1.4	2.5	0.5	65
SA-335 P91, HP-15 SHTR 1 to HP SHTR 2	3			0	0	1.4	1.495	0	0
1.25″ WT, SA-335 P91, HP-16 SHTR 1 to HP SHTR 2	10	4	5	40	2.2	1.4	2.8	0.5	279
1.032″ WT, SA-335 P91, HP-16 SHTR 1 to HP SHTR 2	8	2		16	1.45	1.4	2.5	0.5	65
SA-335 P91, HP-16 SHTR 1 to HP SHTR 2	3	0	0	0	1.20	1.4	1.495		0
1.25″ WT, SA-335 P91, HP-17 SHTR 1 to HP SHTR 2	10	4	5	40	2.2	1.4	2.8	0.5	279
1.031″ WT, SA-335 P91, HP-17 SHTR 1 to HP SHTR 2	8	2		16	1.45	1.4	2.5	0.5	65
SA-335 P91, HP-17 SHTR 1 to HP SHTR 2	3			0			1.495		0
Column total		55	54	398					1467

Facility—Combined cycle power plant
Equipment—Triple-pressure w/Reheat for F-Class GT-Three Wide

4.5.4   Estimate sheet 3—Boltup of flanged joint by weight class

Description	Size	150#/ 300# Boltup	600#/ 900# Boltup	1500#/ 2500# Boltup	DI	MH/ DI	MH/ DI	MH/ DI	PF MH
HP piping and supports									
HP-03 Econ 1 to HP Econ 2	8	0	0	0	0	0.40	0.50	0.65	0
HP-03 Econ 1 to HP Econ 2	6	0	0	0	0	0.40	0.50	0.65	0
HP-04 Econ 2 to HP steam drum	8	0	0	0	0	0.40	0.50	0.65	0
HP-04 Econ 2 to HP steam drum	6	0	0	0	0	0.40	0.50	0.65	0
HP-09 steam drum to HP SH 1	6	0	0	0	0	0.40	0.50	0.65	0
HP-10 steam drum to HP SH 1	6	0	0	0	0	0.40	0.50	0.65	0
HP-11 steam drum to HP SH 1	6	0	0	0	0	0.40	0.50	0.65	0
HP-12 steam drum to HP SH 1	6	0	0	0	0	0.40	0.50	0.65	0
HP-13 steam drum to HP SH 1	6	0	0	0	0	0.40	0.50	0.65	0
HP-14 steam drum to HP SH 1	6	0	0	0	0	0.40	0.50	0.65	0
HP-15 SHTR 1 to HP SHTR 2	10	0	0	0	0	0.40	0.50	0.65	0
HP-15 SHTR 1 to HP SHTR 2	8	0	0	0	0	0.40	0.50	0.65	0
HP-15 SHTR 1 to HP SHTR 2	3	0	0	0	0	0.40	0.50	0.65	0

(Continued)

(Continued)

Description	Size	150#/ 300# Boltup	600#/ 900# Boltup	1500#/ 2500# Boltup	DI	MH/ DI	MH/ DI	MH/ DI	PF MH
HP-16 SHTR 1 to HP SHTR 2	10	0	0	0	0	0.40	0.50	0.65	0
HP-16 SHTR 1 to HP SHTR 2	8	0	0	0	0	0.40	0.50	0.65	0
HP-16 SHTR 1 to HP SHTR 2	3	0	0	0	0	0.40	0.50	0.65	0
HP-17 SHTR 1 to HP SHTR 2	10	0	0	0	0	0.40	0.50	0.65	0
HP-17 SHTR 1 to HP SHTR 2	8	0	0	0	0	0.40	0.50	0.65	0
HP-17 SHTR 1 to HP SHTR 2	3	0	0	0	0	0.40	0.50	0.65	0
Column total					0				0

Facility—Combined cycle power plant
Equipment—Triple-pressure w/Reheat for F-Class GT-Three Wide

4.5.5 Estimate sheet 4—Handle valves by weight class

HP piping and supports	Size	150#/ 300# Valve	600#/ 900# Valve	1500#/ 2500# Valve	DI	MH/ DI	MH/ DI	MH/ DI	PF MH
HP-03 Econ 1 to HP Econ 2									
HP-03 Econ 1 to HP Econ 2	8	0	0	0	0	0.45	0.90	1.8	0
HP-04 Econ 2 to HP steam drum	6	0	0	0	0	0.45	0.90	1.8	0

(Continued)

(Continued)

Description		150#/ 300#	600#/ 900#	1500#/ 2500#					PF
HP piping and supports	Size	Valve	Valve	Valve	DI	MH/ DI	MH/ DI	MH/ DI	MH
HP-04 Econ 2 to HP steam drum	8	0	0	0	0	0.45	0.90	1.8	0
HP-09 steam drum to HP SH 1	6	0	0	0	0	0.45	0.90	1.8	0
HP-10 steam drum to HP SH 1	6	0	0	0	0	0.45	0.90	1.8	0
HP-11 steam drum to HP SH 1	6	0	0	0	0	0.45	0.90	1.8	0
HP-12 steam drum to HP SH 1	6	0	0	0	0	0.45	0.90	1.8	0
HP-13 steam drum to HP SH 1	6	0	0	0	0	0.45	0.90	1.8	0
HP-14 steam drum to HP SH 1	6	0	0	0	0	0.45	0.90	1.8	0
HP-15 SHTR 1 to HP SHTR 2	6	0	0	0	0	0.45	0.90	1.8	0
HP-15 SHTR 1 to HP SHTR 2	10	0	0	0	0	0.45	0.90	1.8	0
HP-15 SHTR 1 to HP SHTR 2	8	0	0	0	0	0.45	0.90	1.8	0
HP-16 SHTR 1 to HP SHTR 2	3	0	0	0	0	0.45	0.90	1.8	0
HP-16 SHTR 1 to HP SHTR 2	10	0	0	0	0	0.45	0.90	1.8	0
HP-16 SHTR 1 to HP SHTR 2	8	0	0	0	0	0.45	0.90	1.8	0
HP-17 SHTR 1 to HP SHTR 2	3	0	0	0	0	0.45	0.90	1.8	0

(Continued)

(Continued)

Description		150#/ 300#	600#/ 900#	1500#/ 2500#					PF
HP piping and supports	Size	Valve	Valve	Valve	DI	MH/ DI	MH/ DI	MH/ DI	MH
HP-17 SHTR 1 to HP SHTR 2	10	0	0	0	0	0.45	0.90	1.8	0
HP-17 SHTR 1 to HP SHTR 2	8	0	0	0	0	0.45	0.90	1.8	0
	3	0	0	0	0	0.45	0.90	1.8	0
Column total					0				0

Facility—Combined cycle power plant
Equipment—Triple-pressure w/Reheat for F-Class GT-Three Wide

4.5.6 Estimate sheet 5—Pipe supports

				Pipe			PF
Description	Material	Size	Sch/Thk	Support	DI	MH/DI	MH
HP piping and supports							
HP-03 Econ 1 to HP Econ 2	SA-106-B	8	140/.906		0	1.00	0
HP-03 Econ 1 to HP Econ 2	SA-106-B	6	160/.906		0	1.00	0
HP-04 Econ 2 to HP steam drum	SA-106-B	8	140/.812	2	16	1.00	16
HP-04 Econ 2 to HP steam drum	SA-106-C	6	160/.906		0	1.00	0
HP-09 steam drum to HP SH 1	SA-106-B	6	160/.906	1	6	1.00	6
HP-10 steam drum to HP SH 1	SA-106-B	6	160/.906	1	6	1.00	6
HP-11 steam drum to HP SH 1	SA-106-B	6	160/.906	1	6	1.00	6
HP-12 steam drum to HP SH 1	SA-106-B	6	160/.906	1	6	1.00	6
HP-13 steam drum to HP SH 1	SA-106-B	6	160/.906	1	6	1.00	6
HP-14 steam drum to HP SH 1	SA-106-B	6	160/.906	1	6	1.00	6
HP-15 SHTR 1 to HP SHTR 2	SA-335-P91	10	1.25	2	20	1.00	20

(*Continued*)

(Continued)

Description	Material	Size	Sch/Thk	Pipe Support	DI	MH/DI	PF MH
HP-15 SHTR 1 to HP SHTR 2	SA-335-P91	8	1.031		0	1.00	0
HP-15 SHTR 1 to HP SHTR 2	SA-335-P91	3			0	1.00	0
HP-16 SHTR 1 to HP SHTR 2	SA-335-P91	10	1.25		0	1.00	0
HP-16 SHTR 1 to HP SHTR 2	SA-335-P91	8	1.031		0	1.00	0
HP-16 SHTR 1 to HP SHTR 2	SA-335-P91	3			0	1.00	0
HP-17 SHTR 1 to HP SHTR 2	SA-335-P91	10	1.25	2	20	1.00	20
HP-17 SHTR 1 to HP SHTR 2	SA-335-P91	8	1.031		0	1.00	0
HP-17 SHTR 1 to HP SHTR 2	SA-335-P91	3			0	1.00	0
Column total				12	92		92

Facility—Combined cycle power plant
Equipment—Triple-pressure w/Reheat for F-Class GT-Three Wide

4.5.7 Estimate sheet 6—Instrument

Description	Material	Size	Sch/The	Instrument	MH/ea.	PF MH
HP piping and supports						
HP-03 Econ 1 to HP Econ 2	SA-106-B	8	140/.906		1.20	0
HP-03 Econ 1 to HP Econ 2	SA-106-B	6	160/.906	9	1.20	10.8
HP-04 Econ 2 to HP steam drum	SA-106-B	8	140/.812	6	1.20	7.2
HP-04 Econ 2 to HP steam drum	SA-106-C	6	160/.906		1.20	0
HP-09 steam drum to HP SH 1	SA-106-B	6	160/.906	4	1.20	4.8
HP-10 steam drum to HP SH 1	SA-106-B	6	160/.906	4	1.20	4.8
HP-11 steam drum to HP SH 1	SA-106-B	6	160/.906	4	1.20	4.8
HP-12 steam drum to HP SH 1	SA-106-B	6	160/.906	4	1.20	4.8

(Continued)

(Continued)

Description	Material	Size	Sch/The	Instrument	MH/ea.	PF MH
HP-13 steam drum to HP SH 1	SA-106-B	6	160/.906	4	1.20	4.8
HP-14 steam drum to HP SH 1	SA-106-B	6	160/.906	4	1.20	4.8
HP-15 SHTR 1 to HP SHTR 2	SA-335-P91	10	1.25	5	1.20	6
HP-15 SHTR 1 to HP SHTR 2	SA-335-P91	8	1.031		1.20	0
HP-15 SHTR 1 to HP SHTR 2	SA-335-P91	3			1.20	0
HP-16 SHTR 1 to HP SHTR 2	SA-335-P91	10	1.25	5	1.20	6
HP-16 SHTR 1 to HP SHTR 2	SA-335-P91	8	1.031		1.20	0
HP-16 SHTR 1 to HP SHTR 2	SA-335-P91	3			1.20	0
HP-17 SHTR 1 to HP SHTR 2	SA-335-P91	10	1.25	5	1.20	6
HP-17 SHTR 1 to HP SHTR 2	SA-335-P91	8	1.031		1.20	0
HP-17 SHTR 1 to HP SHTR 2	SA-335-P91	3			1.20	0
Column total				54		64.8

Facility—Combined cycle power plant
Equipment—Triple-pressure w/Reheat for F-Class GT-Three Wide

4.5.8 Estimate sheet 7—Summary HP piping and supports

Description	PF MH	MH/lf
Estimate sheet 1—Handle and install pipe-welded joint	1558	
Estimate sheet 2—Welding: BW, SW, PWHT arc-uphill	1467	
Estimate sheet 3—Boltup of flanged joint by weight class	0	
Estimate sheet 4—Handle valves by weight class	0	
Estimate sheet 5—Pipe supports	92	
Estimate sheet 6—Instrument	65	
Column total	3182	6.1

4.6   Piping summary converted to MH/lf

HP piping and supports	Pipe	Pipe Hdl	Welding	PS	Instrument	MH	MH/ lf
HP-03 Econ 1 to HP Econ 2	25	55	0	0	0	55	2.2
HP-03 Econ 1 to HP Econ 2	42	61	102	0	10.8	173	4.1
HP-04 Econ 2 to HP steam drum	96	211	61	16	7.2	295	3.1
HP-04 Econ 2 to HP steam drum	1	1	30	0	0	31	31.1
HP-09 steam drum to HP SH 1	24	35	45	6	4.8	91	3.8
HP-10 steam drum to HP SH 1	23	33	35	6	4.8	79	3.4
HP-11 steam drum to HP SH 1	23	33	45	6	4.8	89	3.9
HP-12 steam drum to HP SH 1	23	33	35	6	4.8	79	3.4
HP-13 steam drum to HP SH 1	23	33	45	6	4.8	89	3.9
HP-14 steam drum to HP SH 1	23	33	35	6	4.8	79	3.4
HP-15 SHTR 1 to HP SHTR 2	68	333	279	20	6	638	9.4
HP-15 SHTR 1 to HP SHTR 2	3	9	65	0	0	75	24.9
HP-15 SHTR 1 to HP SHTR 2	1	1	0	0	0	1	0.6
HP-16 SHTR 1 to HP SHTR 2	68	333	279	0	6	618	9.1
HP-16 SHTR 1 to HP SHTR 2	3	9	65	0	0	75	24.9
HP-16 SHTR 1 to HP SHTR 2	1	1	0	0	0	1	0.6
HP-17 SHTR 1 to HP SHTR 2	68	333	279	20	6	638	9.4
HP-17 SHTR 1 to HP SHTR 2	3	9	65	0	0	75	24.9
HP-17 SHTR 1 to HP SHTR 2	1	1	0	0	0	1	0.6
Column total	519	1558	1467	92	65	3182	6.1

Chapter 5

Computer-aided estimation

5.1 Introduction

Computerized estimating is a standard tool in the construction industry to automate and control estimating spreadsheets used to estimate more effectively and efficiently. Estimates are prepared with greater accuracy, more rapidly, and with less effort. The computer stores, retrieves, and provides history of comparable historical data for civil, structural steel, piping, process equipment, pipeline, and storage tanks to produce detailed process plant estimates. The computer when given historical data, from a database of previous similar work task, can reliably calculate material cost and labor by category and rapidly produce both summary and detailed man-hour and cost estimates.

Engineers and estimators use Microsoft Excel spreadsheets to calculate craft labor. The qualitative techniques are primarily based on comparisons of similarities between new and actual field data. Quantitative techniques are based on detailed analysis of historical data and construction activities. Quantitative technique is activity-based costing using formulas to calculate the amount of time it takes to do a task, referred to as cycle time. The estimator must make the right assumptions to derive the cycle time for a task. Level of expertise of the estimator has an influence on the accuracy of the cost estimate.

5.2 Benefits of computer-aided estimating

- System effectively provides a history of labor hours and cost.
- System has accuracy of estimates and consistency between estimators.
- It provides history of comparable historical data.
- System draws from the database.
- Database for worker labor and wage rates, equipment rental, and material cost.
- Process plant construction database for work scope and labor hours.
- Everyone in the company can access and use the database.

Industrial Construction Estimating Manual. DOI: https://doi.org/10.1016/B978-0-12-823362-7.00005-3

5.3 Computer Excel estimate spreadsheets

- Step-by-step how to set up and organize estimate spreadsheets
  1. *Quantity takeoff—detailed work-up sheet*
     The quantity takeoff identifies through interpretation of the drawings, specifications, work scope, and erection sequence, the direct craft labor, and material that will be required to complete the project.
  2. Labor rates and cost—calculation of labor cost
     Labor rates and cost—calculation of labor cost direct and indirect craft
  3. *Labor rates and cost—calculation of labor cost craft supervision*
  4. *Labor rates and cost—calculation of labor cost project staff*
  5. *General conditions' (GC's) cost*
     Rates for materials, rentals
     Rates for equipment—company and third party
  6. *Material cost*
  7. *Subcontractor cost*
  8. *Estimate summary*
     Direct craft labor
     Indirect craft labor
     Craft supervision
     Project on-site staff
     Subtotal—labor
     Site GC—materials
     Site GC—rentals
     Equipment rentals—contractor
     Equipment rentals—third party
     Subtotal—equipment and GC
     Project materials
     Sales tax
     Subcontractors
     Special services
     Subtotal—materials and subcontractors
     Subtotal—job cost
     Fee on labor
     Fee on equipment and GCs
     Fee on material and subcontractors
     Total fee
     Subtotal—job cost + fee
     Bond on contractor
     Contingency
     Gross receipts tax

  Quoted price

5.4   Illustration computer Excel estimate spreadsheet forms

5.4.1   Quantity takeoff—detailed work-up sheet 1 (civil)

Description	Historical MH	Qty	Unit	Estimate Qty	Unit	Carpenter	IW	Laborer
		Historical			Estimate			
Scope						0	0	0
				0		0	0	0
				0		0	0	0
				0		0	0	0
	Historical				Estimate			
	Scope					0	0	0
				0		0	0	0
				0		0	0	0
				0		0	0	0
				0		0	0	0
				0		0	0	0
				0		0	0	0
				0		0	0	0
				0		0	0	0
				0		0	0	0

5.4.2   Labor rates and cost—calculation of labor cost sheet 2

Craft	Labor cost—ST Man-hours	Rate ($/h)	Amount	MH	Rate ($/h)	Amount	Labor cost—OT premium Subsistence Rate ($/h)	Amount ($)
Boilermaker	0	0.00	0	0	0.00	0	0.00	0
Ironworker	0	0.00	0	0	0.00	0	0.00	0
Millwright	0	0.00	0	0	0.00	0	0.00	0
Pipefitter	0	0.00	0	0	0.00	0	0.00	0
Carpenter	0	0.00	0	0	0.00	0	0.00	0
Laborer	0	0.00	0	0	0.00	0	0.00	0
Total direct MH	0							
Subtotal		0				0		0

Travel			
Show up time	0	$0.00	$0
Labor escalation			
Total to summary			**$0**

Indirect craft

| Activity | Labor cost—ST | | | Labor cost—OT premium | | Subsistence | | | |
	Man-hours	Rate ($/h)	Amount ($)	MH	Rate ($/h)	Amount ($)	Rate ($/h)	Amount ($)
Mobe/demobe	0	0.00	0	0	0.00	0	0.00	
Cleanup laborer	0	0.00	0	0	0.00	0	0.00	
Crane operator	0	0.00	0	0	0.00	0	0.00	
Teamster—warehouse	0	0.00	0	0	0.00	0	0.00	
Truck driver	0	0.00	0	0	0.00	0	0.00	
Firewatch laborer	0	0.00	0	0	0.00	0	0.00	
Fire blanketing laborer	0	0.00	0	0	0.00	0	0.00	
Safety training	0	0.00	0	0	0.00	0	0.00	
Welder certification	0	0.00	0	0	0.00	0	0.00	
Safety drug testing	0	0.00	0	0	0.00	0	0.00	
Total indirect MH	0			0				
Subtotal			0				0	0

Travel			
Show up time	0	$0.00	$0
Labor escalation			
Total to summary			**$0**

5.4.3 Labor rates and cost—calculation of labor cost craft supervision sheet 3

Craft supervision labor and cost

		Man-hours	
Craft	Direct MH	Foreman	GF
Boilermaker	0	0	
Ironworker	0	0	
Millwright	0	0	
Pipefitter	0	0	0
Carpenter	0	0	0
Laborer	0	0	

Craft	Labor cost—ST			Labor cost—OT premium			Subsistence	
	MH	ST rate ($/h)	Amount ($)	MH	OT rate ($/h)	Amount ($)	Rate ($/h)	Amount ($)
Boilermaker GF	0	0.00	0	0	0.00	0	**0.00**	0
Boilermaker F	0	0.00	0	0	0.00	0	**0.00**	0
Ironworker GF	0	0.00	0	0	0.00	0	**0.00**	0
Ironworker F	0	0.00	0	0	0.00	0	**0.00**	0
Millwright GF	0	0.00	0	0	0.00	0	**0.00**	0
Millwright F	0	0.00	0	0	0.00	0	**0.00**	0
Pipefitter GF	0	0.00	0	0	0.00	0	**0.00**	0
Pipefitter F	0	0.00	0	0	0.00	0	**0.00**	0
Carpenter GF	0	0.00	0	0	0.00	0	**0.00**	0
Carpenter F	0	0.00	0	0	0.00	0	**0.00**	0
Laborer F	0	0.00	0	0	0.00	0	**0.00**	0
Total craft supervision MH	0							
Subtotal			0			0		0
Total craft supervision labor		0						
Total OT premium			0					
Subsistence			**0**					
Travel	Calculate		**0**					
Labor escalation	Calculate		**0**					
Show up time	Calculate		**0**					
Total to summary			0					

5.4.4   Labor rates and cost—calculation of labor cost project staff sheet 4

Salaried	Labor cost—ST			Labor cost—OT premium			Subsistence	
	Man-hours	ST rate ($/h)	Amount ($)	MH	OT rate ($/h)	Amount ($)	Rate ($/h)	Amount ($)
Construction manager	0	0.00	0	**0.00**		0	0.00	0
Project engineer	0	0.00	0	**0.00**		0	0.00	0
Field engineer	0	0.00	0	**0.00**		0	0.00	0
Safety/QC engineer	0	0.00	0	**0.00**		0	0.00	0
Scheduler	0	0.00	0	**0.00**		0	0.00	0
Quality control	0	0.00	0	**0.00**		0	0.00	0
Office manager	0	0.00	0	**0.00**		0	0.00	0
Cost engineer	0	0.00	0	**0.00**		0	0.00	0
Secretarial	0	0.00	0	**0.00**		0	0.00	0
Material control	0	0.00	0	**0.00**		0	0.00	0
Purchasing	0	0.00	0	**0.00**		0	0.00	0
	0	0.00	0	**0.00**		0	0.00	0

(*Continued*)

(Continued)

Project on-site staff	Labor cost—ST			Labor cost—OT premium		Subsistence	
Salaried	Man-hours	ST rate ($/h)	Amount ($)	MH OT rate ($/h)	Amount ($)	Rate ($/h)	Amount ($)
General superintendent							
Total staff MH	0						
Subtotal			0		0		0
Total staff MH			0				
Total OT premium			0				
Subsistence			0				
Travel	**Calculate**		**0**				
Home—office travel	**Calculate**		**0**				
Staff housing	**Calculate**		**0**				
Total to summary			0				

5.4.5 General conditions' cost sheet 5

Site general conditions	Materials				Rentals		
Description	Duration	Rate ($)	Cost ($)	Description	Duration	Rate ($)	Cost ($)
Electrical—hookup	0	0	0	Office trailer	0.0	0	0
Electrical—power/ monthly	0	0	0	Office trailer—setup	0	0	0
Telephone—hookup	0	0	0	Change trailers	0.0	0	0
Telephone—monthly	0.0	0	0	Change trailers—setup	0	0	0
Office furniture	0.0	0	0	Tool trailers	0.0	0	0
Office supplies	0.0	0	0	Sanitary facilities	0	0	0
Fax machine	0	0	0	Dumpsters	0.0	0	0
Copier	0	0	0	Temporary fencing		0	0
Drinking water	0.0	0	0	Security		0	0
Plan reproduction	0	0	0	Cargo containers	0	0	0
Project safety billboard	0	0	0	Construction water hookup	0	0	0
Permits	0	0	0	Temporary air/water piping	0	0	0
Total to summary			0	Total to summary			0

Equipment rentals

Company owned				Third party			
Description	Duration	Rate ($)	Cost ($)	Description	Duration	Rate ($)	Cost ($)
3/4 ton pickup	0	0	0	Manitowoc 2250/Maxer w/ 220' Boom	0	0	0
Job site pickup	0.0	0	0	Assembly/mobe/tear down/ demobe	0	0	0
2 ton flatbed truck	0	0	0	200 ton crawler crane w/ 210' Boom	0.0	0	0
Forklift, over 8000#	0	0	0	Assembly/mobe/tear down/ demobe	0	0	0
Reachlift,8000#−10,000#	0.0	0	0	90 ton rough terrain crane	0	0	0
Hydraulic crane, 10 ton	0	0	0	Mobe/demobe	0	0	0
Hydraulic crane, 60 ton	0	0	0	60 ton rough terrain crane	0	0	0
Hydraulic crane, 80 ton	0	0	0	Mobe/demobe	0	0	0
Truck crane, 8 ton	0	0	0	Engineering (HRSG erection)	0	0	0
Electric welder, 8 Pak	0	0	0	Ten (10) 9 ton deadman—5 loads	0	0	0
Industrial welding generator	0.0	0	0	Freight per load	0	0	0
Air compressor, 175−475 CFM	0	0	0	Crane mats (per sf)	0	0	0
Generator, 60−100 kW	0.0	0	0	Freight per load	0	0	0
Hydrostatic test pump, 4 GPM	0.0	0	0	Technician	0	0	0
Fusion machine, 6"−8"	0	0	0	Crane fuel and maintenance	0.0	0	0
Fusion machine, 10"−18"		0	0	500 ton gantry	0	0	0
Trailer, Pipe Dolly	0.0	0	0	Mobe/demobe	0	0	0
Zoom Boom, 60'	0	0	0	Prime mover	0	0	0
Zoom Boom, 40'	0	0	0	Mobe/demobe	0	0	0
Van, 8 Passenger		0	0	Barge ramp		0	0
Light tower	0	0	0	Mobe/demobe		0	0
Total to summary			0	Total to summary			0

5.4.6   Material cost sheet 6

Project material

Description	Qty	$/Qty	Amount ($)
Weld test coupons	0	0	0
Safety items	0	0	0
Fire blankets	0	0	0
Nomex suits	0	0	0
Steel toe shoes	0	0	0
Fuel for equipment	0	0	0
Special tools	0	0	0
H2S monitor	0	0	0

(Continued)

(Continued)

Description	Qty	$/Qty	Amount ($)
Pipe supports	0	0	0
Bolt and gasket sets	0	0	0
Plate	0	0	0
Grout	0	0	0
Fill oil	0	0	0
			0
Subtotal			0
Escalation	**0.00%**	%	0
Freight	**0.00%**	%	0
Subtotal			0
Sales tax	0.00%	%	0
		Material/sales tax	0

5.4.7 Subcontractor cost sheet 7

Subcontractors

Subcontractor	Type	MH	Amount ($)	Service	Vendor	Amount ($)
TBD	Scaffold	**0**		Deputy inspector—civil		
	Insulation			Deputy inspector—structural		
TBD	Painting	**0**		Welding inspector		
	Electrical			Weld testing Concrete testing		
TBD	NDE	**0**		Welder certification lab fees		
	Lead abatement			Outside engineering		
				Survey and layout	**TBD**	**0**
				Third-party inspector	**TBD**	**0**
				Soil monitoring/ testing		
				Hot tapping		**0**
	PWHT	**0**		Drug testing laboratory	**TBD**	**0**
TBD	Millwright	**0**				

(*Continued*)

(Continued)

Subcontractor	Type	MH	Amount ($)	Service	Vendor	Amount ($)
Total to summary			0	Total to summary		0

5.4.8 Estimate summary sheet 8

Estimate summary

Account description	Man-hours	Cost ($)	Adjustment ($)	Final cost ($)
Direct craft labor	0	0	0	0
Indirect craft labor	0	0	0	0
Union rebate				0
Craft supervision	0	0	0	0
Project on-site Staff	0	0	0	0
Subtotal—labor	**0**	**0**	**0**	**0**
Site GC—materials		0	0	0
Site GC—Rentals		0	0	0
Equipment rentals—		0	0	0
Equipment rentals—third party		0	0	0
Subtotal—equipment and GC		**0**	**0**	**0**
Project materials		Amount ($)		
Sales tax		0	0	0
Subcontractors		0	0	0
Special services		0	0	0
Subtotal—materials and subcontractors		**0**	**0**	**0**
Subtotal—job cost		**0**	**0**	**0**
Fee on labor	0	%		0
Fee on equipment and GCs	0	%		0
Fee on material and subcontractors	0	%		0
Total fee				0
Subtotal—job cost + fee				**0**
Bond on ARB	0.00	%		0
Contingency				
Gross receipts tax	0.00	%		0
Quoted price			**0**	**0**

5.5 Sample cost estimate: Simple cycle power plant SCR Foundation estimate

5.5.1 Quantity takeoff detailed work-up sheet

| Install drilled piers and place SCR Foundation estimate sheet 1 | | | Historical | | Estimate | | | |

Description	MH	Qty	Unit Qty	Unit	Carpenter	IW	Laborer
		Historical		Estimate			
Drilled piers					**568**	**528**	**0**
2' drilled piers	480.00	1.00	job	1.00 ea	480		0
Fabricate and install rebar cage	24.00	22.00	ea	22.00 ea		528	
4000# PSI concrete piers 8 year/day/ea	4.00	22.00	ea	22.00 ea	88		
		Historical		Estimate			
SCR Foundation (30′−0″ × 40′−0″)					**407**	**393**	**1025**
Structure excavation—SCR Foundation	0.60	177.78	cy	177.78 cy		107	
Fabricate, install, strip foundation forms (one use)	0.35	140.00	lf	140.00 lf	49		
Fabricate and install rebar	65.00	4.40	ton	4.40 ton		286	
Place concrete slab	2.50	250.00	cy	250.00 ea			625
Layout, templates, and set anchor bolts	1.20	88.00	ea	88.00 ea	106		
Set embedded steel	0.06	500.00	lb	500.00 lb	30		
Finish flat concrete	0.15	1200.00	sf	1200.00 sf	180		
Structure backfill and compact—loader and Wacker	0.60	71.11	cy	71.11 cy	43		
Cleanup and grade area w/skip loader	320.00	1.00	job	1.00 job			320
Haul spoils to location	80.00	1.00	job	1.00 job			80

	Actual		Estimated			
Facility—simple cycle power plant	MH	Carpenter	IW		Laborer	MH
Drilled Piers	1020	568	528		0	1096
SCR Foundation (30′−0″ × 40′−0″)	1820	407	393		1025	1825

5.5.2 Labor rates and cost—calculation of labor cost direct and indirect craft sheet 2

Estimate no.	Start date	
Project	Completion date	
Direct craft	Calendar days	**64**

Craft	Labor cost—ST			Labor cost—OT premium			Subsistence	
	Man-hours	Rate ($/h)	Amount ($)	Man-hours	Rate ($/h)	Amount ($)	Rate ($/h)	Amount ($)
Carpenter	975	78.56	76,617	0	32.80	0	0.00	0
Ironworker	921	84.94	78,201	0	28.84	0		0
Millwright	0	0.00	0	0	0.00	0		0
Pipefitter	0	0.00	0	0	0.00	0		0
Carpenter	0	0.00	0	0	0.00	0		0
Laborer	1025	68.84	70,561	0	26.45	0		0
Total direct MH	2921							
Subtotal			225,379				0	0
Travel								
Show up time	0	0.00	0					
Labor escalation								
Total to summary			**225,379**					

Indirect craft

Activity	Labor cost—ST			Labor cost—OT premium			Subsistence	
	Man-hours	Rate ($/h)	Amount ($)	Man-hours	Rate ($/h)	Amount ($)	Rate ($/h)	Amount ($)
Mobe/demobe	320	85.11	27,234		33.75	0		
Cleanup laborer	0	0.00	0		0.00	0		
Crane operator	320	93.21	29,827		33.40	0		
Oiler	0	0.00	0		0.00	0		
Truck driver	80	71.54	5723		23.19	0		
Firewatch—laborer	0	0.00	0		0.00	0		
Fire blanketing laborer	0	0.00	0		0.00	0		
Safety training	20	85.11	1702					
Welder certification	0	0.00	0		0.00	0		
Safety drug testing	20	85.11	1702		33.75	0		
Total indirect MH	760							
Subtotal			66,188				0	0
Travel								
Show up time	0	85.11	0					
Labor escalation								
Total to summary			**66,188**					

5.5.3 Labor rates and cost—calculation of labor cost craft supervision sheet 3

Project		
Supervision		Calendar days **64**
Total work days	46	
Hours per work day	8	
Months	2.1	

Craft supervision man-hours

	Man-hours		
Craft	Direct MH	Foreman	Gen Foreman
Carpenter	975	0	368
Ironworker	921	120	0
Millwright	0	0	0
Pipefitter	0	0	0
Carpenter	0	0	0
Laborer	1025	128	0

	Labor cost—ST			Labor cost—OT premium			Subsistence	
Craft	Man-hours	Rate ($/h)	Amount ($)	Man-hours	Rate ($/h)	Amount ($)	Rate ($/h)	Amount ($)
Boilermaker GF	0	0.00	0		0.00	0	0.00	0
Boilermaker F	0	0.00	0		0.00	0	0.00	0
Ironworker GF	0	0.00	0		0.00	0		0
Ironworker F	120	90.14	10,817		31.45	0		0
Millwright GF	0	0.00	0		0.00	0		0
Millwright F	0	0.00	0		0.00	0		0
Pipefitter GF	0	0.00	0	0	0.00	0		0
Pipefitter F	0	0.00	0	0	0.00	0		0
Carpenter GF	368	86.96	32,001		37.00	0		0
Carpenter F	0	0.00	0		0.00	0		0
Laborer F	128	77.59	9932		30.82	0		0
Total craft supervision MH	616							
Subtotal			52,750			0		0
Total craft supervision labor		52,750						
Total OT premium			0					
Subsistence			0					
Travel	Calculate		0					
Labor escalation	Calculate		0					
Show up time	Calculate		0					
Total to summary			52,750					

5.5.4 Labor rates and cost—calculation of labor cost project staff sheet 4

Estimate no.		Start date	
Project		Completion date	
Project on-site staff		Calendar days	64
Total work days	46		
Hours per work day	8		
Total work hours	366		
Months	2.1		

	Labor cost—ST			Labor cost—OT premium			Subsistence	
Salaried	Man-hours	Rate ($/H)	Amount ($)	Man-hours	Rate ($/H)	Amount ($)	Rate ($/H)	Amount ($)
Construction manager	366	135.00	49,371		135.00	0	16.50	6034
Project engineer	0	0.00	0		0.00	0	0.00	0
Field engineer	183	80.00	14,629		80.00	0	0.00	0
Safety/QC engineer	0	0.00	0		0.00	0	0.00	0
Scheduler	0	0.00	0		0.00	0	0.00	0
Quality control	0	0.00	0		0.00	0	0.00	0
Office manager	0	0.00	0		0.00	0	0.00	0
Cost engineer	0	0.00	0		0.00	0	0.00	0
Secretarial	366	35.00	12,800		35.00	0	0.00	0
Material control	0	0.00	0		0.00	0	0.00	0
Purchasing	0	0.00	0		0.00	0	0.00	0
General superintendent	0	0.00	0		0.00	0	0.00	0
Total staff MH	914							
Subtotal			76,800			0		6034
Total staff MH			76,800					
Total OT premium			0					
Subsistence			6034					
Travel	Calculate		0					
Home—office travel	Calculate		0					
Staff housing	Calculate		0					

5.5.5 General conditions' cost sheet 5

Estimate no.		Start date	
Project		Completion date	
Site GC		Calendar days	64
Total work days	46		
Months	2.1		

	Materials				Rentals		
Description	Duration	Rate ($)	Cost ($)	Description	Duration	Rate ($)	Cost ($)
Electrical—hookup	0	0.00	0	Office trailer	2.1	1000	2133
Electrical—power/ monthly	0	0.00	0	Office trailer—setup	1	250	250
Telephone—hookup	0	0.00	0	Change trailers	2.1	200	427
Telephone—monthly	0.0	0.00	0	Change trailers—setup	1	250	250
Office furniture	2.1	1500	3200	Tool trailers	2.1	200	427
Office supplies	2.1	2000	4267	Sanitary facilities	4	200	800
Fax machine	0	0.00	0	Dumpsters	2.1	1500	3200
Copier	0	0.00	0	Temporary fencing			0
Drinking water	2.1	2500	5333	Security			0
Plan reproduction	0	0.00	0	Cargo containers	0	0.00	0
Project safety billboard	0	0.00	0	Construction water hookup	1	5000	5000
Permits	0	0.00	0	Temporary air/water piping	0	0.00	0
Total to summary			12,800	Total to summary			12,487

Equipment rentals

	Company owned				Third party		
Description	Duration	Rate ($)	Cost ($)	Description	Duration	Rate ($)	Cost ($)
3/4 ton pickup	2.1	2838	6054	Manitowoc 2250/Maxer w/ 220′ Boom	0	0.00	0
Job site pickup	2.1	2838	6054	Assembly/mobe/tear down/ demobe	0	0.00	0
2 ton flatbed truck	0	0.00	0	200 ton crawler crane w/ 210′ Boom	0.0	0.00	0
Forklift, over 8000#	0	0.00	0	Assembly/mobe/tear down/ demobe	0	0.00	0
Reachlift,8000#−10,000#	0.0	4000	0	90 ton rough terrain crane	0	0.00	0
Hydraulic crane, 10 ton		7482	0	Mobe/demobe	0	0.00	0
Hydraulic crane, 60 ton		17,200	0	60 ton rough terrain crane	0	0.00	0
Hydraulic crane, 80 ton		21,500	0	Mobe/demobe	0	0.00	0
Truck crane, 8 ton	0	0.00	0	Engineering (HRSG erection)	0	0.00	0

(*Continued*)

(Continued)

Company owned				Third party			
Description	Duration	Rate ($)	Cost ($)	Description	Duration	Rate ($)	Cost ($)
Electric welder, 8 pack	0	0.00	0	Ten (10) 9 ton deadman—5 loads	0	0.00	0
Industrial welding generator	0.0	0.00	0	Freight per load	0	0.00	0
Air compressor, 175–475 CFM	2.1	900	1920	Crane mats (per sf)	0	0.00	0
Generator, 60–100 kW		0.00	0	Freight per load	0	0.00	0
Hydrostatic test pump, 4 GPM	0.0	1000	0	Bigge Technician	0	0.00	0
Cat Backhoe 305	2.1	5676	12,109	Crane fuel and maintenance	0.0	0.00	0
Dump truck 10 year/day	0.47	8686	4040	500 ton gantry	0	0.00	0
Trailer, Pipe Dolly	0.0	0.00	0	Mobe/demobe	0	0.00	0
Mini Excavator 305	1.1	5676	5960	Prime mover	0	0.00	0
Compactor	1.1	1462	1608	Mobe/demobe	0	0.00	0
Concrete vibrator	2.1	430	903	Barge ramp		0.00	0
Light tower	0	0.00	0	Mobe/demobe		0.00	0
Total to summary			38,649	Total to summary			0

5.5.6 Material cost sheet 6

Estimate no.		Start date	
		Completion date	
		Calendar days	64
Total work days	46		
Months	2.1		
Project material			

Description	Qty	$/Qty	Amount ($)
Weld test coupons	0	0	0
Safety items:	8	0	0
Fire blankets	0	0	0
Nomex Suits	0	0	0
Steel toe shoes	8	0	0
Fuel for equipment	0	0	0
Special tools	0	0	0
Concrete (4000# PSI)	426	124	52,824
Anchor bolts/rebar (lb)	3408	9	28,968
	0	0	0
Piping	0	0	0
Grout, steel, etc.	0	0	0
Oil fill	0		0
Miscellaneous	1	0	0

(Continued)

(Continued)

Description	Qty	$/Qty	Amount ($)
			0
Subtotal			81,792
Escalation	0.00%	%	0
Freight	0.00%	%	0
Subtotal			81,792
California sales tax	9.75%	%	7975

5.5.7 Subcontractor cost sheet 7

Estimate no.				Start date		
Project				Completion date		
Total work days	46			Calendar days	64	
Months	2.1					
Subcontractors				**Special services**		

Subcontractor	Type	MH	Amount ($)	Service	Vendor	Amount ($)
	Scaffold	0	0	Deputy inspector—civil		
	Insulation			Deputy inspector—structural		
Redwood	Painting		0	Welding inspector		
	Electrical			Weld testing Concrete testing		
TBD	NDE		0	Welder certification lab fees		
PCI	Lead abatement			Outside engineering		
				Survey and layout	1800	18,000
				Third-party inspector	0	0
				Soil monitoring/testing		
				Hot tapping		0
Superheat	PWHT			Drug testing laboratory	0	0

(*Continued*)

(Continued)

Subcontractor	Type	MH	Amount ($)	Service	Vendor	Amount ($)
	Supervisor 2- to 6-way consoles	0	0			
Lube oil flush Drilling contractor	To be determined		20,000			
Total to summary			20,000	Total to summary		18,000

5.5.8 Estimate summary sheet 8

Estimate no.	Start date	
Bid date	Completion date	
	Calendar days	64
	Total work days	46
Estimate summary		

Account description	Man-hours	Cost ($)	Adjustment ($)	Final cost ($)
Direct craft labor	2921	225,379	0	225,379
Indirect craft labor	760	66,188	0	66,188
Union rebate				0
Craft supervision	616	52,750	0	52,750
Project on-site staff	914	82,834	0	82,834
Subtotal—labor	**5211**	**427,151**	**0**	**427,151**

Site GC—materials ($)	12,800	0	12,800
Site GC—rentals ($)	12,487	0	12,487
Equipment rentals—ARB Inc. ($)	38,649	0	38,649
Equipment rentals—third-party ($)	0	0	0
Special services ($)		0	0
Subtotal—equipment and GC ($)	**63,935**	**0**	**63,935**

Project materials ($)	81,792	0	81,792
Sales tax ($)	7975	0	7975
Subcontractors ($)	20,000	0	20,000
Special services ($)	18,000	0	18,000
Subtotal—materials and subcontractors ($)	**127,767**	**0**	**127,767**
Subtotal—job cost ($)	**618,853**	**0**	**618,853**

Fee on labor	25.00%	%		$106,788
Fee on equipment and GCs	10.00%	%		$6,394
Fee on material and subcontractors	0.00%	%		$0
Total fee	15.46%			$113,181
Subtotal—job cost + fee				**$732,035**
Bond on ARB	0.00%	%		$0
Contingency	10.00%			**$73,203**
Gross receipts tax	0.00%	%		$0
Quoted price			$154.52	**$805,238**

5.5.9 SCR Foundation—bid breakdown sheet 9

Item Description	Unit	$/MH	Man-hours	Labor ($)	Material ($)	Total Dollars ($)	Comments
1.00 Drilled piers	LS		1096	89,470		89,470	
2.00 SCR Foundation (30′–0″ × 40′–0″)	LS		1825	135,909		135,909	
16.00 Craft supervision	LS		616	52,750		52,750	
17.00 Material	LS				81,792	81,792	
18.00 Sub-bond costs	LS					0	
19.00 Sales tax	LS				7975	7975	
20.00 Construction equipment rentals	LS				63,935	63,935	
21.00 Subcontractors	LS				38,000	38,000	
22.00 Overtime/shift work/ productivity	MH					0	
23.00 Indirect and project field staff	MH		1674	149,022		149,022	
24.00 Contingency, home office (G&A), and profit	LS				186,385	186,385	
25.00 Total direct labor man-hours	MH		2921				
26.00 Averaged burdened labor rate	$/MH	$154.52					
			5211	**613,536**	**191,702**	**805,238**	

5.5.10 Estimate analysis sheet 10

Estimate no.	
Bid date	
Calendar days	64
Total work days	46

Project crew, supervision, and staff	Man-hours	Manpower
Direct craft	2921	8
Indirect craft	760	2
Craft supervision	616	2
Staff	914	3
Total site man power	5211	15

Ratio analysis

Supervision percent of craft labor (%)	21.09
Staff percent of craft labor (%)	31.30
Indirect craft percent of craft labor (%)	26.02

Fee loading

Fee on labor (%)	25.00
Fee percent of cost (%)	18.29
Fee percent of price (%)	14.06

Piping analysis

Pipe labor MH/lf	0.00
Pipe welding MH/DI	0.00
Supports percent of craft labor	0.00
Testing percent of craft labor	0.00

SCR Foundation

Scope of work		8	8	64	46	2921	9.1	64	2.1
					Total				
Direct craft	Total MH	Manpower	h/ day	h/ day	Work days	Man-hours	Weeks	Days	Months
	2921	**8**	**8**	**64**	**46**	**2921**	**9.1**	**64**	**2.1**
Drilled piers	1096	8	8	64	17	1096	3.4	24	0.8
SCR Foundation (30′−0″ × 40′−0″)	1825	8	8	64	29	1825	5.7	40	1.3

Chapter 6

Combined cycle power plant (1 × 1) labor estimate

6.1 Introduction

Natural gas combined cycle power plants play an important role in the construction industry. These power plants are the most efficient power plants operating on the power grid with an efficiency between 45% and 57%. Natural gas power plants are attractive to the power industry because of the following advantages:

- It has efficiency between 45% and 57%, highest efficiency for a power plant type.
- It has a small footprint compared to other power plant types.
- Construction time is short compared to other types of power plants.
- Capital cost is lower than other types of power plants.
- Gaseous emissions are very low; plants emit far less carbon dioxide than other types of power plants.

This chapter focuses on detailed labor estimates to determine direct craft man-hours to erect a 1 × 1 combined cycle power plant. The man-hour estimate provides the basis for the project schedule. The estimator uses the labor estimate to determine crew craft and make up requirements, project duration and to develop Level I and II schedules for the project.

6.2 Detailed estimating unit-quantity model

Combined cycle power plant estimate consists of the following work estimates:

- Foundations
- Mechanical equipment [CTG, STG, heat recovery steam generator (HRSG)] includes vendor piping
- BOP equipment
- Structural steel
- Underground piping
- Aboveground piping

Industrial Construction Estimating Manual. DOI: https://doi.org/10.1016/B978-0-12-823362-7.00006-5

6.3 Combine cycle power plant foundation summary

	Carpenter	Laborer	IW	Total MH
6.3.1 Estimate summary—power island equipment foundations	**28,741.0**	**5102.6**	**17,350.9**	**6287.4**	**28,741.0**
Boiler foundations	**7347.1**	**1137.7**	**4550.4**	**1659.0**	**7347.1**
Estimate sheet—HRSG and stack foundation	6276.9	774.0	4004.9	1498.0	6276.9
Estimate sheet—HRSG RH blowdown tank foundation	93.6	42.0	44.6	7.0	93.6
Estimate sheet—CEMS building foundation	112.8	48.0	53.6	11.2	112.8
Estimate sheet—steam cycle chemistry feed area foundation	249.6	84.2	129.0	36.4	249.6
Estimate sheet—boiler feedwater pump foundation (2 ea.)	270.9	81.7	138.9	50.4	270.9
Estimate sheet—auxiliary boiler foundation	343.3	107.9	179.4	56.0	343.3
CTG and CTG-related foundations	**5528.5**	**1025.5**	**3329.8**	**1173.2**	**5528.5**
Estimate sheet—combustion turbine and CTG generator	4280.5	600.2	2686.2	994.0	4280.5
Estimate sheet—generator circuit breaker foundation (2 ea.)	321.1	74.25	188.05	58.8	321.1
Estimate sheet—fuel gas performance heater foundation	297.7	80.7	163.8	53.2	297.7
Estimate sheet—rotor air cooler foundation	138.6	58.8	63.0	16.8	138.6
Estimate sheet—excitation skid and transformer foundation	144.2	58.8	67.15	18.2	144.2
Estimate sheet—VT and surge cubicle foundation	69.7	31.2	32.85	5.6	69.7
Estimate sheet—sample panel foundation	116.6	50.4	53.61	12.6	116.6
Estimate sheet—compressor wash skid foundation	102.2	44.1	48.31	9.8	102.2
Estimate sheet—GT water wash tank foundation	58.1	27	26.85	4.2	58.1

(Continued)

(Continued)

	Carpenter	Laborer	IW	Total MH
STG and STG-related foundations	**15,865.3**	**2939.5**	**9470.7**	**3455.2**	**15,865.3**
Estimate—steam turbine and steam turbine generator	8498.4	1043.5	5424.9	2030.0	8498.4
Estimate—closed cooling water heat exchangers (2 ea.)	215.8	92.25	102.5	21	215.8
Estimate—closed cooling water pumps (2 ea.)	186.0	87.3	87.5	11.2	186.0
Estimate—cooling tower chemical feed area	468.5	111.75	265.7	91	468.5
Estimate—circulating water pump station and cooling tower	582.6	142.2	328.4	112	582.6
Estimate—circulating water pump station and cooling tower	5196.2	1200	2912.6	1084	5196.2
Estimate—sulfuric acid, coagulant tank, and pump foundations	390.2	114	206.15	70	390.2
Estimate—condenser polishing unit	103.4	48	44.15	11.2	103.4
Estimate—condenser extraction pumps	91.5	45	38.1	8.4	91.5
Estimate—condenser vacuum pumps	133.0	55.5	60.7	16.8	133.0
Plant mechanical system–related foundations	**7160.7**	**2046.3**	**3962.2**	**1152.2**	**7160.7**
Estimate sheet— potable water skid	132.3	54	64.3	14	132.3
Estimate sheet— demineralized storage tank	297.8	87	160.35	50.4	297.8
Estimate sheet—oil–water separator (process drain) foundation	91.9	42	42.9	7	91.9
Estimate sheet—oil–water separator (storm drain) foundation	144.5	54.45	70.4	19.6	144.5
Estimate sheet— demineralized water forwarding pump skid	64.4	29.7	33.25	1.4	64.4
Estimate sheet—injection water storage tank	323.9	88.5	185.0	50.4	323.9

(Continued)

(Continued)

	Carpenter	Laborer	IW	Total MH
Estimate sheet—STIG wastewater injection pump skid	64.4	29.7	33.3	1.4	64.4
Estimate sheet—clarifier (2 ea.)	381.0	135	192.8	53.2	381.0
Estimate sheet—clarified water storage tank	2552.6	322.5	1630.9	599.2	2552.6
Estimate sheet—clarified water forwarding pump	104.6	45	49.8	9.8	104.6
Estimate sheet—gas compressors (2 ea.)	492.7	168	250.45	74.2	492.7
Estimate sheet—STIG plant gas compressor	260.0	88.8	130.55	40.6	260.0
Estimate sheet—gas compressor filter foundation	136.0	54.75	65.8	15.4	136.0
Estimate sheet—fuel gas filter separator foundation	75.8	33.6	36.6	5.6	75.8
Estimate sheet—control oil skid foundation	86.3	36.6	41.25	8.4	86.3
Estimate sheet—STG plant heat exchanger	195.9	63.3	101.8	30.8	195.9
Estimate sheet—dry lime silos (5 ea.)	477.4	210	228.2	39.2	477.4
Estimate sheet—air compressors (2 ea.)	175.9	79.5	83.8	12.6	175.9
Estimate sheet—compressed air dryer	77.5	36.0	37.3	4.2	77.5
Estimate sheet—compressed air receiver	77.5	36.0	37.3	4.2	77.5
Estimate sheet—clarified water and solids holding tanks (3 ea.)	452.1	170.1	226.0	56.0	452.1
Estimate sheet—waste solids and recycle solids pump skid (2 ea.)	219.7	58.5	123.4	37.8	219.7
Estimate sheet—clarified water pump skid foundation	104.6	45.0	49.8	9.8	104.6
Estimate sheet—eyewash, showers, heaters (3 ea.)	108.1	48.6	53.9	5.6	108.1

6.3.2 Utility distribution plant electrical controls foundations	17,628.3	5932.4	8138.5	3557.4	17,628.3
Excel estimate HRSG utility bridge foundations (4 ea.)	1777.8	504.0	961.6	312.2	1777.8
Excel estimate STG utility bridge foundations (4 ea.)	1315.0	453.6	622.0	239.4	1315.0
Excel estimate stair towers and ladders (4 ea.)	199.3	96.0	97.7	5.6	199.3
Excel estimate stair towers and ladders (4 ea.)	130.0	57.6	65.4	7.0	130.0
Excel estimate miscellaneous pipe support foundation (100 ea.)	416.7	288.0	86.7	42.0	416.7
Excel estimate miscellaneous concrete paving (100 ea.)	253.9	32.4	172.5	49.0	253.9
Excel estimate CTG iso-phase bus duct Foundation (16 ea.)	265.6	120.0	106.4	39.2	265.6
Excel estimate STG iso-phase bus duct foundation (16 ea.)	265.6	120.0	106.4	39.2	265.6
Excel estimate CTG step-up transformer foundation (16 ea.)	1322.2	453.6	605.4	263.2	1322.2
Excel estimate STG main step-up transformer foundation (16 ea.)	1322.2	453.6	605.4	263.2	1322.2
Excel estimate auxiliary transformer (16 ea.)	1303.3	645.0	446.9	211.4	1303.3
Excel estimate SUS transformer (5 ea.)	404.6	202.5	168.5	33.6	404.6
Excel estimate CTG PDC	429.5	133.5	206.4	89.6	429.5
Excel estimate STG/HRSG PDC	488.1	126.0	251.5	110.6	488.1
Excel estimate miscellaneous cable tray supports and fdn. (5 ea.)	247.4	192.0	45.6	9.8	247.4
Excel estimate transmission towers (6 ea.)	412.2	210.6	151.2	50.4	412.2
Excel estimate takeoff towers (7 ea.)	406.4	239.4	123.6	43.4	406.4
Excel estimate disconnect switch	131.2	51.6	62.8	16.8	131.2
Excel estimate circuit breaker	143.5	63.0	67.9	12.6	143.5
Excel estimate pole supports (14 ea.)	136.2	75.6	46.6	14.0	136.2
Excel estimate miscellaneous small foundations (10 ea.)	298.8	216.0	64.6	18.2	298.8
Excel estimate electrical manholes (7 ea.)	1679.2	861.8	562.6	254.8	1679.2
Excel estimate transmission towers (6 ea.)	412.2	210.6	151.2	50.4	412.2
Excel estimate transmission towers (6 ea.)	846.1	126.0	360.3	359.8	846.1
Excel estimate trenches, curbs, equip pads, piers, etc.	3021.8	0.0	1999.8	1022.0	3021.8

6.4 Work estimates are illustrated with the unit-quantity model

6.4.1 Estimate—power island equipment foundations

Excel estimate spreadsheet for boiler foundations

Description	Historical		Unit	Estimate		Carpenter	Labor	IW
	MH	Qty		Qty				
	MH_a	n_i		n_i	$n_i MH_a$			
Estimate sheet—HRSG and stack foundation					6276.9	774.0	4004.9	1498.0
Excavation (bathtub)	0.400	0.0	cy	0.0	0.0			
Forms (204' × 54' × 5')	0.300	2580	sf	2580.0	774.0	774.0		
Rebar	14.0	107	ton	107.0	1498.0			1498.0
Concrete (200' × 50' × 5')	2.0	1944.4	cy	1944.4	3888.9		3888.9	
Mud mat (204' × 54' × 0.33')	0.500	136.0	cy	136.0	68.0		68.0	
Backfill, compact	0.600	80.0	cy	80.0	48.0		48.0	
Estimate sheet—HRSG RH blowdown tank foundation					93.6	42.0	44.6	7.0
Excavation (bathtub)	0.400	0.0	cy	0.0	0.0			
Forms (16' × 16' × 2.5')	0.300	140	sf	140.0	42.0	42.0		
Rebar	14.0	0.5	ton	0.5	7.0			7.0
Concrete (10' × 10' × 2.5')	2.0	9.7	cy	9.7	19.4		19.4	
Mud mat (14' × 14' × 0.33')	0.500	2.4	cy	2.4	1.2		1.2	
Backfill, compact	0.600	40.0	cy	40.0	24.0		24.0	
Estimate sheet—CEMS building foundation					112.8	48.0	53.6	11.2
Excavation (bathtub)	0.400	0.0	cy	0.0	0.0			
Forms (14' × 14' × 2.5')	0.300	160	sf	160.0	48.0	48.0		
Rebar	14.0	0.8	ton	0.8	11.2			11.2
Concrete (12' × 12' × 2.5')	2.0	14.0	cy	14.0	28.0		28.0	
Mud mat (16' × 16' × 0.33')	0.500	3.2	cy	3.2	1.6		1.6	
Backfill, compact	0.600	40.0	cy	40.0	24.0		24.0	
Estimate sheet—steam cycle chemistry feed area foundation					249.6	84.2	129.0	36.4
Excavation (bathtub)	0.400	0.0	cy	0.0	0.0			
Forms (42' × 19' × 3.3')	0.300	280.6	sf	280.6	84.2	84.2		
Rebar	14.0	2.6	ton	2.6	36.4			36.4
Concrete (38' × 15' × 2.3')	2.0	51.0	cy	51.0	102.0		102.0	
Mud mat (42' × 19' × 0.33')	0.500	9.9	cy	9.9	5.0		5.0	
Backfill, compact	0.600	36.8	cy	36.8	22.1		22.1	
Estimate sheet—boiler feedwater pump foundation (2 ea.)					270.9	81.7	138.9	50.4
Excavation (bathtub)	0.400	0.0	cy	0.0	0.0			
Forms (26.75' × 15.25' × 2.5')	0.300	272.2	sf	272.2	81.7	81.7		
Rebar	14.0	3.6	ton	3.6	50.4			50.4

(*Continued*)

(Continued)

Description	Historical MH	Qty	Unit	Estimate Qty		Carpenter	Labor	IW
	MH_a	n_i		n_i	$n_i\,MH_a$			
Estimate sheet—HRSG					6276.9	774.0	4004.9	1498.0
and stack foundation								
Concrete	2.0	52.4	cy	52.4	104.7		104.7	
(22.75′ × 11.25′ × 3.3′)								
Mud mat	0.500	5.0	cy	5.0	2.5		2.5	
(26.75′ × 15.25′ × 0.33′)								
Backfill, compact	0.600	52.8	cy	52.8	31.7		31.7	
Estimate sheet—auxiliary					**343.3**	**107.9**	**179.4**	**56.0**
boiler foundation								
Excavation (bathtub)	0.400	0.0	cy	0.0	0.0			
Forms (34′ × 24′ × 3.1′)	0.300	359.6	sf	359.6	107.9	107.9		
Rebar	14.0	4	ton	4.0	56.0			56.0
Concrete (30′ × 20′ × 3.1′)	2.0	72.3	cy	72.3	144.6		144.6	
Mud mat (34′ × 24′ × 0.33′)	0.500	10.1	cy	10.1	5.1		5.1	
Backfill, compact	0.600	49.6	cy	49.6	29.8		29.8	

6.4.2 CTG and CTG-related foundations sheet 1

Description	Historical MH	Qty	Unit	Estimate Qty		Carpenter	Labor	IW
	MH_a	n_i		n_i	$n_i\,MH_a$			
Estimate sheet—					4280.5	600.2	2686.2	994.0
combustion turbine								
and CTG generator								
Excavation (bathtub)	0.400	0.0	cy	0.0	0.0			
Forms (57′ × 107′ × 6.1′)	0.300	2000.8	sf	2000.8	600.2	600.2		
Rebar	14.0	71	ton	71.0	994.0			994.0
Concrete (53′ × 103′ × 6.1′)	2.0	1295.0	cy	1295.0	2590.0		2590.0	
Mud mat	0.500	75.3	cy	75.3	37.7		37.7	
(57′ × 107′ × 0.33′)								
Backfill, compact	0.600	97.6	cy	97.6	58.6		58.6	
Estimate sheet—generator					**321.1**	**74.3**	**188.1**	**58.8**
circuit breaker foundation								
(2 ea.)								
Excavation (bathtub)	0.400	0.0	cy	0.0	0.0			
Forms (29′ × 20.5′ × 2.5′)	0.300	247.5	sf	247.5	74.3	74.3		
Rebar	14.0	4.2	ton	4.2	58.8			58.8
Concrete	2.0	80.2	cy	80.2	160.4		160.4	
(25′ × 16.5′ × 2.5′)								
Mud mat	0.500	7.3	cy	7.3	3.7		3.7	
(29′ × 20.5′ × 0.33′)								
Backfill, compact	0.600	40.0	cy	40.0	24.0		24.0	
					297.7	**80.7**	**163.8**	**53.2**

(*Continued*)

(Continued)

Description	Historical MH	Qty	Unit	Estimate Qty		Carpenter	Labor	IW
	MH$_a$	n$_i$		n$_i$	n$_i$ MH$_a$			
Estimate sheet— combustion turbine and CTG generator					4280.5	600.2	2686.2	994.0
Estimate sheet—fuel gas performance heater foundation								
Excavation (bathtub)	0.400	0.0	cy	0.0	0.0			
Forms (57' × 24.5' × 1.7')	0.300	268.9	sf	268.9	80.7	80.7		
Rebar	14.0	3.8	ton	3.8	53.2			53.2
Concrete (53' × 20.5' × 1.7')	2.0	69.7	cy	69.7	139.4		139.4	
Mud mat (57' × 24.5' × 0.33')	0.500	17.2	cy	17.2	8.6		8.6	
Backfill, compact	0.600	26.4	cy	26.4	15.8		15.8	
Estimate sheet—rotor air cooler foundation					138.6	58.8	63.0	16.8
Excavation (bathtub)	0.400	0.0	cy	0.0	0.0			
Forms (33' × 12' × 2')	0.300	196	sf	196.0	58.8	58.8		
Rebar	14.0	1.2	ton	1.2	16.8			16.8
Concrete (33' × 8' × 2')	2.0	20.5	cy	20.5	41.0		41.0	
Mud mat (37' × 12' × 0.33')	0.500	5.5	cy	5.5	2.8		2.8	
Backfill, compact	0.600	32.0	cy	32.0	19.2		19.2	
Estimate sheet—excitation skid and transformer foundation					144.2	58.8	67.2	18.2
Excavation (bathtub)	0.400	0.0	cy	0.0	0.0			
Forms (26.33' × 17' × 2')	0.300	196	sf	196.0	58.8	58.8		
Rebar	14.0	1.3	ton	1.3	18.2			18.2
Concrete (22.3' × 13' × 2')	2.0	22.6	cy	22.6	45.2		45.2	
Mud mat (26.33' × 17' × 0.33')	0.500	5.5	cy	5.5	2.8		2.8	
Backfill, compact	0.600	32.0	cy	32.0	19.2		19.2	

6.4.3 CTG and CTG-related foundations sheet 2

Description	Historical MH	Qty	Unit	Estimate Qty		Carpenter	Labor	IW
	MH$_a$	n$_i$		n$_i$	n$_i$ MH$_a$			
Estimate sheet—VT and surge cubicle foundation					69.7	31.2	32.9	5.6
Excavation (bathtub)	0.400	0.0	cy	0.0	0.0			
Forms (13.5' × 12.5' × 2')	0.300	104	sf	104.0	31.2	31.2		
Rebar	14.0	0.4	ton	0.4	5.6			5.6

(Continued)

(Continued)

Description	Historical			Estimate		Carpenter	Labor	IW
	MH	Qty	Unit	Qty				
	MH$_a$	n$_i$		n$_i$	n$_i$ MH$_a$			
Estimate sheet—VT and surge cubicle foundation					69.7	31.2	32.9	5.6
Concrete (9.5′ × 8.5′ × 2′)	2.0	6.3	cy	6.3	12.6		12.6	
Mud mat (13.5′ × 12.5′ × 0.33′)	0.500	2.1	cy	2.1	1.1		1.1	
Backfill, compact	0.600	.32.0	cy	32.0	19.2		19.2	
Estimate sheet—sample panel foundation					116.6	50.4	53.6	12.6
Excavation (bathtub)	0.400	0.0	cy	0.0	0.0			
Forms (38′ × 12′ × 2.1′)	0.300	168	sf	168.0	50.4	50.4		
Rebar	14.0	0.9	ton	0.9	12.6			12.6
Concrete (24′ × 8′ × 2.1′)	2.0	15.7	cy	15.7	31.4		31.4	
Mud mat (28′ × 12′ × 0.33′)	0.500	4.1	cy	4.1	2.1		2.1	
Backfill, compact	0.600	33.6	cy	33.6	20.2		20.2	
Estimate sheet—compressor wash skid foundation					102.2	44.1	48.3	9.8
Excavation (bathtub)	0.400	0.0	cy	0.0	0.0			
Forms (22′ × 13′ × 2.1′)	0.300	147	sf	147.0	44.1	44.1		
Rebar	14.0	0.7	ton	0.7	9.8			9.8
Concrete (18′ × 9′ × 2.1′)	2.0	13.2	cy	13.2	26.4		26.4	
Mud mat (22′ × 13′ × 0.33′)	0.500	3.5	cy	3.5	1.8		1.8	
Backfill, compact	0.600	33.6	cy	33.6	20.2		20.2	
Estimate sheet—GT water wash tank foundation					58.1	27.0	26.9	4.2
Excavation (bathtub)	0.400	0.0	cy	0.0	0.0			
Forms (20′ × 10′ × 1.5′)	0.300	90	sf	90.0	27.0	27.0		
Rebar	14.0	0.3	ton	0.3	4.2			4.2
Concrete (16′ × 6′ × 1.5′)	2.0	5.6	cy	5.6	11.2		11.2	
Mud mat (20′ × 10′ × 0.33′)	0.500	2.5	cy	2.5	1.3		1.3	
Backfill, compact	0.600	24.0	cy	24.0	14.4		14.4	

6.4.4 STG and STG-related foundations sheet 3

Description	Historical			Estimate		Carpenter	Labor	IW
	MH	Qty	Unit	Qty				
	MH$_a$	n$_i$		n$_i$	n$_i$ MH$_a$			
Estimate—steam turbine and steam turbine generator					8498.4	1043.5	5424.9	2030.0
Excavation (bathtub)	0.400	0.0	cy	0.0	0.0			
Forms (120′ × 67′ × 9.3′)	0.300	3478.2	sf	3478.2	1043.5	1043.5		
Rebar	14.0	145	ton	145.0	2030.0			2030.0
Concrete (116′ × 63′ × 9.3′)	2.0	2643.0	cy	2643.0	5286.0		5286.0	

(Continued)

(Continued)

Description	Historical MH	Qty	Unit	Estimate Qty		Carpenter	Labor	IW
	MH_a	n_i		n_i	$n_i MH_a$			
Estimate—steam turbine and steam turbine generator					8498.4	1043.5	5424.9	2030.0
Mud mat (120' × 67' × 0.33')	0.500	99.3	cy	99.3	49.7		49.7	
Backfill, compact	0.600	148.8	cy	148.8	89.3		89.3	
Estimate—closed cooling water heat exchangers (2 ea.)					**215.8**	**92.3**	**102.5**	**21.0**
Excavation (bathtub)	0.400	0.0	cy	0.0	0.0			
Forms (17' × 13.75' × 2.5')	0.300	307.5	sf	307.5	92.3	92.3		
Rebar	14.0	1.5	ton	1.5	21.0			21.0
Concrete (13' × 9.75' × 2.5')	2.0	25.8	cy	25.8	51.6		51.6	
Mud mat (17' × 13.75' × 0.33')	0.500	5.8	cy	5.8	2.9		2.9	
Backfill, compact	0.600	80.0	cy	80.0	48.0		48.0	
Estimate—closed cooling water pumps (2 ea.)					**186.0**	**87.3**	**87.5**	**11.2**
Excavation (bathtub)	0.400	0.0	cy	0.0	0.0			
Forms (15' × 9' × 3')	0.300	291	sf	291.0	87.3	87.3		
Rebar	14.0	0.8	ton	0.8	11.2			11.2
Concrete (11.25' × 5' × 3')	2.0	14.1	cy	14.1	28.2		28.2	
Mud mat (15' × 9' × 0.33')	0.500	3.4	cy	3.4	1.7		1.7	
Backfill, compact	0.600	96.0	cy	96.0	57.6		57.6	
Estimate—cooling tower chemical feed area					**468.5**	**111.8**	**265.7**	**91.0**
Excavation (bathtub)	0.400	0.0	cy	0.0	0.0			
Forms (41' × 34' × 2.5')	0.300	372.5	sf	372.5	111.8	111.8		
Rebar	14.0	6.5	ton	6.5	91.0			91.0
Concrete (36.5' × 30' × 2.5')	2.0	116.6	cy	116.6	233.2		233.2	
Mud mat (41' × 34' × 0.33')	0.500	17.0	cy	17.0	8.5		8.5	
Backfill, compact	0.600	40.0	cy	40.0	24.0		24.0	
Estimate—circulating water pump station and cooling tower					**582.6**	**142.2**	**328.4**	**112.0**
Excavation (bathtub)	0.400	0.0	cy	0.0	0.0			
Forms (48' × 31' × 3')	0.300	474	sf	474.0	142.2	142.2		
Rebar	14.0	8	ton	8.0	112.0			112.0
Concrete (44' × 27' × 3')	2.0	145.2	cy	145.2	290.4		290.4	
Mud mat (48' × 31' × 0.33')	0.500	18.4	cy	18.4	9.2		9.2	
Backfill, compact	0.600	48.0	cy	48.0	28.8		28.8	

6.4.5 Circulating water pump station and cooling tower

Circulating water pump station and cooling tower					**5196.2**	**1200.0**	**2912.6**	**1083.6**
Excavation (bathtub)	0.400	0.0	cy	0.0	0.0			
Forms (76′ × 124′ × 4′)	0.300	4000	sf	4000.0	1200.0	1200.0		
Rebar	14.0	77.4	ton	77.4	1083.6			1083.6
Concrete (72′ × 120′ × 4′)	2.0	1408.0	cy	1408.0	2816.0		2816.0	
Mud mat (76′ × 124′ × 0.33′)	0.500	116.3	cy	116.3	58.2		58.2	
Backfill, compact	0.600	64.0	cy	64.0	38.4		38.4	
Estimate—sulfuric acid, coagulant tank, and pump foundations					**390.2**	**114.0**	**206.2**	**70.0**
Excavation (bathtub)	0.400	0.0	cy	0.0	0.0			
Forms (55′ × 21′ × 2.5′)	0.300	380	sf	380.0	114.0	114.0		
Rebar	14.0	5	ton	5.0	70.0			70.0
Concrete (51′ × 17′ × 2.5′)	2.0	92.3	cy	92.3	184.6		184.6	
Mud mat (55′ × 21′ × 0.33′)	0.500	14.3	cy	14.3	7.2		7.2	
Backfill, compact	0.600	24.0	cy	24.0	14.4		14.4	
Estimate—condenser polishing unit					**103.4**	**48.0**	**44.2**	**11.2**
Excavation (bathtub)	0.400	0.0	cy	0.0	0.0			
Forms (18′ × 14′ × 2.5′)	0.300	160	sf	160.0	48.0	48.0		
Rebar	14.0	0.8	ton	0.8	11.2			11.2
Concrete (14′ × 10′ × 2.5′)	2.0	14.1	cy	14.1	28.2		28.2	
Mud mat (18′ × 14′ × 0.33′)	0.500	3.1	cy	3.1	1.6		1.6	
Backfill, compact	0.600	24.0	cy	24.0	14.4		14.4	
Estimate—condenser extraction pumps					**91.5**	**45.0**	**38.1**	**8.4**
Excavation (bathtub)	0.400	0.0	cy	0.0	0.0			
Forms (19′ × 11′ × 2.5′)	0.300	150	sf	150.0	45.0	45.0		
Rebar	14.0	0.6	ton	0.6	8.4			8.4
Concrete (15′ × 7′ × 2.5′)	2.0	11.2	cy	11.2	22.4		22.4	
Mud mat (19′ × 11′ × 0.33′)	0.500	2.6	cy	2.6	1.3		1.3	
Backfill, compact	0.600	24.0	cy	24.0	14.4		14.4	
Estimate—condenser vacuum pumps					**133.0**	**55.5**	**60.7**	**16.8**
Excavation (bathtub)	0.400	0.0	cy	0.0	0.0			
Forms (20′ × 17′ × 2.5′)	0.300	185	sf	185.0	55.5	55.5		
Rebar	14.0	1.2	ton	1.2	16.8			16.8
Concrete (16′ × 13′ × 2.5′)	2.0	22.1	cy	22.1	44.2		44.2	
Mud mat (19′ × 17′ × 0.33′)	0.500	4.2	cy	4.2	2.1		2.1	
Backfill, compact	0.600	24.0	cy	24.0	14.4		14.4	

6.4.6 Plant mechanical system—related foundations sheet 1

Description	Historical			Estimate				
	MH	Qty	Unit	Qty		Carpenter	Labor	IW
	MH$_a$	n$_i$		n$_i$	n$_i$ MH$_a$			
Estimate—Plant mechanical system—related foundations					**3165.5**	**355.7**	**2667.0**	**142.8**
Estimate sheet—potable water skid					**132.3**	**54.0**	**64.3**	**14.0**
Excavation (bathtub)	0.400	0.0	cy	0.0	0.0			
Forms (22′ × 14′ × 2.5′)	0.300	180	sf	180.0	54.0	54.0		
Rebar	14.0	1	ton	1.0	14.0			14.0
Concrete (18′ × 10′ × 2.5′)	2.0	19.2	cy	19.2	38.4		38.4	
Mud mat (22′ × 14′ × 0.33′)	0.500	3.8	cy	3.8	1.9		1.9	
Backfill, compact	0.600	40.0	cy	40.0	24.0		24.0	
Estimate sheet—demineralized storage tank					**297.8**	**87.0**	**160.4**	**50.4**
Excavation (bathtub)	0.400	0.0	cy	0.0	0.0			
Forms (32′ × 26′ × 2.5′)	0.300	290	sf	290.0	87.0	87.0		
Rebar	14.0	3.6	ton	3.6	50.4			50.4
Concrete (28′ × 22′ × 2.5′)	2.0	65.6	cy	65.6	131.2		131.2	
Mud mat (32′ × 26′ × 0.33′)	0.500	10.3	cy	10.3	5.2		5.2	
Backfill, compact	0.600	40.0	cy	40.0	24.0		24.0	
Estimate sheet—oil—water separator (process drain) foundation					**91.9**	**42.0**	**42.9**	**7.0**
Excavation (bathtub)	0.400	0.0	cy	0.0	0.0			
Forms (18′ × 10′ × 2.5′)	0.300	140	sf	140.0	42.0	42.0		
Rebar	14.0	0.5	ton	0.5	7.0			7.0
Concrete (14′ × 6′ × 2.5′)	2.0	8.9	cy	8.9	17.8		17.8	
Mud mat (18′ × 10′ × 0.33′)	0.500	2.2	cy	2.2	1.1		1.1	
Backfill, compact	0.600	40.0	cy	40.0	24.0		24.0	
Estimate sheet—oil—water separator (storm drain) foundation					**144.5**	**54.5**	**70.4**	**19.6**
Excavation (bathtub)	0.400	0.0	cy	0.0	0.0			
Forms (47 × 14′ × 1.5′)	0.300	181.5	sf	181.5	54.5	54.5		
Rebar	14.0	1.4	ton	1.4	19.6			19.6
Concrete (42.5′ × 10′ × 1.5′)	2.0	26.0	cy	26.0	52.0		52.0	
Mud mat (47′ × 14′ × 0.33′)	0.500	8.0	cy	8.0	4.0		4.0	
Backfill, compact	0.600	24.0	cy	24.0	14.4		14.4	
Estimate sheet—demineralized water forwarding pump skid					**64.4**	**29.7**	**33.3**	**1.4**
Excavation (bathtub)	0.400	0.0	cy	0.0	0.0			
Forms (6.5 × 10′ × 3′)	0.300	99	sf	99.0	29.7	29.7		
Rebar	14.0	0.1	ton	0.1	1.4			1.4
Concrete (2.5′ × 6′ × 3′)	2.0	1.8	cy	1.8	3.6		3.6	

(*Continued*)

(Continued)

Description	Historical MH	Qty	Unit	Estimate Qty		Carpenter	Labor	IW
	MH_a	n_i		n_i	$n_i\,MH_a$			
Mud mat (6.5' × 10' × 0.33')	0.500	1.7	cy	1.7	0.9		0.9	
Backfill, compact	0.600	48.0	cy	48.0	28.8		28.8	
Estimate sheet—injection water storage tank					**323.9**	**88.5**	**185.0**	**50.4**
Excavation (bathtub)	0.400	0.0	cy	0.0	0.0			
Forms (32 × 27' × 2.5')	0.300	295	sf	295.0	88.5	88.5		
Rebar	14.0	3.6	ton	3.6	50.4			50.4
Concrete (28' × 23' × 2.5')	2.0	65.6	cy	65.6	131.2		131.2	
Mud mat (32' × 27' × 0.33')	0.500	59.6	cy	59.6	29.8		29.8	
Backfill, compact	0.600	40.0	cy	40.0	24.0		24.0	

6.4.7 Plant mechanical system—related foundations sheet 2

	Historical			Estimate				
Estimate sheet—injection water forwarding pump						**64.4**	**29.7**	**33.3**	**1.4**
Excavation (bathtub)	0.400	0.0	cy	0.0	0.0			
Forms (6.5 × 10' × 3')	0.300	99	sf	99.0	29.7	29.7		
Rebar	14.0	0.1	ton	0.1	1.4			1.4
Concrete (2.5' × 6' × 3')	2.0	1.8	cy	1.8	3.6		3.6	
Mud mat (6.5' × 10' × 0.33')	0.500	1.7	cy	1.7	0.9		0.9	
Backfill, compact	0.600	48.0	cy	48.0	28.8		28.8	
Estimate sheet—STIG wastewater injection pump skid						**64.4**	**29.7**	**33.3**	**1.4**
Excavation (bathtub)	0.400	0.0	cy	0.0	0.0			
Forms (6.5 × 10' × 3')	0.300	99	sf	99.0	29.7	29.7		
Rebar	14.0	0.1	ton	0.1	1.4			1.4
Concrete (2.5' × 6' × 3')	2.0	1.8	cy	1.8	3.6		3.6	
Mud mat (6.5' × 10' × 0.33')	0.500	1.7	cy	1.7	0.9		0.9	
Backfill, compact	0.600	48.0	cy	48.0	28.8		28.8	
Estimate sheet—clarifier (2 ea.)						**381.0**	**135.0**	**192.8**	**53.2**
Excavation (bathtub)	0.400	0.0	cy	0.0	0.0			
Forms (24 × 21' × 2.5')	0.300	450	sf	450.0	135.0	135.0		
Rebar	14.0	3.8	ton	3.8	53.2			53.2
Concrete (20' × 17' × 2.5')	2.0	69.3	cy	69.3	138.6		138.6	
Mud mat (24' × 21' × 0.33')	0.500	12.4	cy	12.4	6.2		6.2	
Backfill, compact	0.600	80.0	cy	80.0	48.0		48.0	

(*Continued*)

(Continued)

	Historical			Estimate				
Estimate sheet—clarified water storage tank					**2552.6**	**322.5**	**1630.9**	**599.2**
Excavation (bathtub)	0.400	0.0	cy	0.0	0.0			
Forms (52 × 163′ × 2.5′)	0.300	1075	sf	1075.0	322.5	322.5		
Rebar	14.0	42.8	ton	42.8	599.2			599.2
Concrete (48′ × 159′ × 2.5′)	2.0	777.3	cy	777.3	1554.6		1554.6	
Mud mat (52′ × 163′ × 0.33′)	0.500	104.6	cy	104.6	52.3		52.3	
Backfill, compact	0.600	40.0	cy	40.0	24.0		24.0	
Estimate sheet—clarified water forwarding pump					**104.6**	**45.0**	**49.8**	**9.8**
Excavation (bathtub)	0.400	0.0	cy	0.0	0.0			
Forms (16 × 14′ × 2.5′)	0.300	150	sf	150.0	45.0	45.0		
Rebar	14.0	0.7	ton	0.7	9.8			9.8
Concrete (12′ × 10′ × 2.5′)	2.0	12.2	cy	12.2	24.4		24.4	
Mud mat (16′ × 14′ × 0.33′)	0.500	2.8	cy	2.8	1.4		1.4	
Backfill, compact	0.600	40.0	cy	40.0	24.0		24.0	
Estimate sheet—gas compressors (2 ea.)					**492.7**	**168.0**	**250.5**	**74.2**
Excavation (bathtub)	0.400	0.0	cy	0.0	0.0			
Forms (38 × 18′ × 2.5′)	0.300	560	sf	560.0	168.0	168.0		
Rebar	14.0	5.3	ton	5.3	74.2			74.2
Concrete (34′ × 14′ × 2.5′)	2.0	97.0	cy	97.0	194.0		194.0	
Mud mat (38′ × 18′ × 0.33′)	0.500	16.9	cy	16.9	8.5		8.5	
Backfill, compact	0.600	80.0	cy	80.0	48.0		48.0	

6.4.8 Plant mechanical system—related foundations sheet 3

	Historical			Estimate				
Estimate sheet—STIG plant gas compressor					**260.0**	**88.8**	**130.6**	**40.6**
Excavation (bathtub)	0.400	0.0		cy	0.0	0.0		
Forms (58 × 16′ × 2′)	0.300	296		sf	296.0	88.8	88.8	
Rebar	14.0	2.9		ton	2.9	40.6		40.6
Concrete (54′ × 12′ × 2′)	2.0	52.8		cy	52.8	105.6	105.6	
Mud mat (58′ × 16′ × 0.33′)	0.500	11.5		cy	11.5	5.8	5.8	
Backfill, compact	0.600	32.0		cy	32.0	19.2	19.2	
Estimate sheet—gas compressor filter foundation					**136.0**	**54.8**	**65.8**	**15.4**
Excavation (bathtub)	0.400	0.0		cy	0.0	0.0		
Forms (21 × 15.5′ × 2.5′)	0.300	182.5		sf	182.5	54.8	54.8	
Rebar	14.0	1.1		ton	1.1	15.4		15.4
Concrete (17′ × 11.5′ × 2.5′)	2.0	19.9		cy	19.9	39.8	39.8	
Mud mat (21′ × 15.5′ × 0.33′)	0.500	4.0		cy	4.0	2.0	2.0	
Backfill, compact	0.600	40.0		cy	40.0	24.0	24.0	

(*Continued*)

(Continued)

	Historical			Estimate				
Estimate sheet—fuel gas filter separator foundation					**75.8**	**33.6**	**36.6**	**5.6**
Excavation (bathtub)	0.400	0.0	cy	0.0	0.0			
Forms (15 × 13′ × 2′)	0.300	112	sf	112.0	33.6	33.6		
Rebar	14.0	0.4	ton	0.4	5.6			5.6
Concrete (11′ × 9′ × 2′)	2.0	8.1	cy	8.1	16.2		16.2	
Mud mat (15′ × 13′ × 0.33′)	0.500	2.4	cy	2.4	1.2		1.2	
Backfill, compact	0.600	32.0	cy	32.0	19.2		19.2	
Estimate sheet—control oil skid foundation					**86.3**	**36.6**	**41.3**	**8.4**
Excavation (bathtub)	0.400	0.0	cy	0.0	0.0			
Forms (16 × 14.5′ × 2′)	0.300	122	sf	122.0	36.6	36.6		
Rebar	14.0	0.6	ton	0.6	8.4			8.4
Concrete (12′ × 10.5′ × 2′)	2.0	10.3	cy	10.3	20.6		20.6	
Mud mat (16′ × 14.5′ × 0.33′)	0.500	2.9	cy	2.9	1.5		1.5	
Backfill, compact	0.600	32.0	cy	32.0	19.2		19.2	
Estimate sheet—STG plant heat exchanger					**195.9**	**63.3**	**101.8**	**30.8**
Excavation (bathtub)	0.400	0.0	cy	0.0	0.0			
Forms (31.5 × 21.5′ × 2′)	0.300	211	sf	211.0	63.3	63.3		
Rebar	14.0	2.2	ton	2.2	30.8			30.8
Concrete (27.5′ × 17.5′ × 2′)	2.0	39.2	cy	39.2	78.4		78.4	
Mud mat (31.5′ × 21.5′ × 0.33′)	0.500	8.4	cy	8.4	4.2		4.2	
Backfill, compact	0.600	32.0	cy	32.0	19.2		19.2	
Estimate sheet—dry lime silos (5 ea.)					**477.4**	**210.0**	**228.2**	**39.2**
Excavation (bathtub)	0.400	0.0	cy	0.0	0.0			
Forms (15 × 13′ × 2.5′)	0.300	700	sf	700.0	210.0	210.0		
Rebar	14.0	2.8	ton	2.8	39.2			39.2
Concrete (11′ × 9′ × 2.5′)	2.0	51.1	cy	51.1	102.2		102.2	
Mud mat (15′ × 13′ × 0.33′)	0.500	12.0	cy	12.0	6.0		6.0	
Backfill, compact	0.600	200.0	cy	200.0	120.0		120.0	

6.4.9 Plant mechanical system—related foundations sheet 4

	Historical			Estimate				
Estimate sheet—air compressors (2 ea.)					**175.9**	**79.5**	**83.8**	**12.6**
Excavation (bathtub)	0.400	0.0	cy	0.0	0.0			
Forms (15 × 11.5′ × 2.5′)	0.300	265	sf	265.0	79.5	79.5		
Rebar	14.0	0.9	ton	0.9	12.6			12.6
Concrete (11′ × 7.5′ × 2.5′)	2.0	16.8	cy	16.8	33.6		33.6	
Mud mat (15′ × 11.5′ × 0.33′)	0.500	4.3	cy	4.3	2.2		2.2	
Backfill, compact	0.600	80.0	cy	80.0	48.0		48.0	
Estimate sheet—compressed air dryer					**77.5**	**36.0**	**37.3**	**4.2**
Excavation (bathtub)	0.400	0.0	cy	0.0	0.0			
Forms (12 × 12′ × 2.5′)	0.300	120	sf	120.0	36.0	36.0		
Rebar	14.0	0.3	ton	0.3	4.2			4.2
Concrete (8′ × 8′ × 2.5′)	2.0	6.2	cy	6.2	12.4		12.4	
Mud mat (12′ × 12′ × 0.33′)	0.500	1.8	cy	1.8	0.9		0.9	
Backfill, compact	0.600	40.0	cy	40.0	24.0		24.0	

(Continued)

(Continued)

		Historical		Estimate				
Estimate sheet—compressed air receiver					**78.9**	**36.0**	**37.3**	**5.6**
Excavation (bathtub)	0.400	0.0	cy	0.0	0.0			
Forms (12 × 12′ × 2.5′)	0.300	120	sf	120.0	36.0	36.0		
Rebar	14.0	0.4	ton	0.4	5.6			5.6
Concrete (8′ × 8′ × 2.5′)	2.0	6.2	cy	6.2	12.4		12.4	
Mud mat (12′ × 12′ × 0.33′)	0.500	1.8	cy	1.8	0.9		0.9	
Backfill, compact	0.600	40.0	cy	40.0	24.0		24.0	
Estimate sheet—clarified water and solids holding tanks (3 ea.)					**452.1**	**170.1**	**226.0**	**56.0**
Excavation (bathtub)	0.400	0.0	cy	0.0	0.0			
Forms (21 × 16.8′ × 2.5′)	0.300	567	sf	567.0	170.1	170.1		
Rebar	14.0	4	ton	4.0	56.0			56.0
Concrete (17′ × 12.8′ × 2.5′)	2.0	73.7	cy	73.7	147.4		147.4	
Mud mat (21′ × 16.8′ × 0.33′)	0.500	13.1	cy	13.1	6.6		6.6	
Backfill, compact	0.600	120.0	cy	120.0	72.0		72.0	
Estimate sheet—waste solids and recycle solids pump skid (2 ea.)					**219.7**	**58.5**	**123.4**	**37.8**
Excavation (bathtub)	0.400	0.0	cy	0.0	0.0			
Forms (21 × 18 × 2.5′)	0.300	195	sf	195.0	58.5	58.5		
Rebar	14.0	2.7	ton	2.7	37.8			37.8
Concrete (17′ × 14′ × 2.5′)	2.0	48.5	cy	48.5	97.0		97.0	
Mud mat (21′ × 18′ × 0.33′)	0.500	4.7	cy	4.7	2.4		2.4	
Backfill, compact	0.600	40.0	cy	40.0	24.0		24.0	
Estimate sheet—clarified water pump skid foundation					**104.6**	**45.0**	**49.8**	**9.8**
Excavation (bathtub)	0.400	0.0	cy	0.0	0.0			
Forms (16 × 14′ × 2.5′)	0.300	150	sf	150.0	45.0	45.0		
Rebar	14.0	0.7	ton	0.7	9.8			9.8
Concrete (12′ × 10′ × 2.5′)	2.0	12.2	cy	12.2	24.4		24.4	
Mud mat (16′ × 14′ × 0.33′)	0.500	2.8	cy	2.8	1.4		1.4	
Backfill, compact	0.600	40.0	cy	40.0	24.0		24.0	
Estimate sheet—eyewash, showers, heaters (3 ea.)					**108.1**	**48.6**	**53.9**	**5.6**
Excavation (bathtub)	0.400	0.0	cy	0.0	0.0			
Forms (9 × 9′ × 1.5′)	0.300	162	sf	162.0	48.6	48.6		
Rebar	14.0	0.4	ton	0.4	5.6			5.6
Concrete (5′ × 5′ × 1.5′)	2.0	4.6	cy	4.6	9.2		9.2	
Mud mat (9′ × 9′ × 0.33′)	0.500	3.0	cy	3.0	1.5		1.5	
Backfill, compact	0.600	72.0	cy	72.0	43.2		43.2	

6.4.10 Utility distribution plant electrical controls sheet 1 foundations

		Historical		Estimate				
Description	MH	Qty	Unit	Qty		Carpenter	Labor	IW
	MH_a	n_i		n_i	$n_i\,MH_a$			
Excel estimate HRSG utility bridge foundations (4 ea.)					**1777.8**	**504.0**	**961.6**	**312.2**
Excavation (bathtub)	0.400	0.0	cy	0.0	0.0			
Forms (29 × 31′ × 3.5′)	0.300	1680	sf	1680.0	504.0	504.0		
Rebar	14.0	22.3	ton	22.3	312.2			312.2
Concrete (25′ × 27′ × 3.5′)	2.0	402.5	cy	402.5	805.0		805.0	
Mud mat (29′ × 31′ × 0.33′)	0.500	44.4	cy	44.4	22.2		22.2	
Backfill, compact	0.600	224.0	cy	224.0	134.4		134.4	
Excel estimate STG utility bridge foundations (4 ea.)					**1315.0**	**453.6**	**622.0**	**239.4**
Excavation (bathtub)	0.400	0.0	cy	0.0	0.0			
Forms (25 × 29′ × 3.5′)	0.300	1512	sf	1512.0	453.6	453.6		
Rebar	14.0	17.1	ton	17.1	239.4			239.4
Concrete (21′ × 25′ × 3.5′)	1.5	313.1	cy	313.1	469.7		469.7	
Mud mat (25′ × 29′ × 0.33′)	0.500	35.8	cy	35.8	17.9		17.9	
Backfill, compact	0.600	224.0	cy	224.0	134.4		134.4	
Excel estimate stair towers and ladders (4 ea.)					**199.3**	**96.0**	**97.7**	**5.6**
Excavation (bathtub)	0.400	0.0	cy	0.0	0.0			
Forms (10 × 10′ × 2′)	0.300	320	sf	320.0	96.0	96.0		
Rebar	14.0	0.4	ton	0.4	5.6			5.6
Concrete (6′ × 6′ × 2′)	1.5	12.3	cy	12.3	18.5		18.5	
Mud mat (10′ × 10′ × 0.33′)	0.500	4.9	cy	4.9	2.5		2.5	
Backfill, compact	0.600	128.0	cy	128.0	76.8		76.8	
Excel estimate stair towers and ladders (4 ea.)					**130.0**	**57.6**	**65.4**	**7.0**
Excavation (bathtub)	0.400	0.0	cy	0.0	0.0			
Forms (8 × 8′ × 1.5′)	0.300	192	sf	192.0	57.6	57.6		
Rebar	14.0	0.5	ton	0.5	7.0			7.0
Concrete (4′ × 4′ × 1.5′)	1.5	4.1	cy	4.1	6.2		6.2	
Mud mat (8′ × 8′ × 0.33′)	0.500	3.2	cy	3.2	1.6		1.6	
Backfill, compact	0.600	96.0	cy	96.0	57.6		57.6	
Excel estimate miscellaneous pipe support fen (100 ea.)					**416.7**	**288.0**	**86.7**	**42.0**
Excavation (bathtub)	0.400	0.0	cy	0.0	0.0			
Forms (4 × 2 × 1.6′)	0.300	960	sf	960.0	288.0	288.0		
Rebar	14.0	3	ton	3.0	42.0			42.0
Concrete (4′ × 2′ × 1.6′)	1.5	54.5	cy	54.5	81.8		81.8	
Mud mat (4′ × 2′ × 0.33′)	0.500	9.9	cy	9.9	5.0		5.0	
Backfill, compact	0.600	0.0	cy	0.0	0.0		0.0	
Excel estimate miscellaneous concrete paving (100 ea.)					**253.9**	**32.4**	**172.5**	**49.0**
Excavation (bathtub)	0.400	0.0	cy	0.0	0.0			
Forms (108 × 4.5 × 0.5′)	0.300	108	sf	108.0	32.4	32.4		

(*Continued*)

(Continued)

Description	MH	Qty	Unit	Qty		Carpenter	Labor	IW
		Historical		Estimate				
	MH_a	n_i		n_i	$n_i\,MH_a$			
Rebar	14.0	3.5	ton	3.5	49.0			49.0
Concrete (108′ × 0.5′ × 0.5′)	1.5	115.0	cy	115.0	172.5		172.5	
Mud mat (108′ × 4.5′ × 0.33′)	0.500	0.0	cy	0.0	0.0		0.0	
Backfill, compact	0.600	0.0	cy	0.0	0.0		0.0	

6.4.11 Utility distribution plant electrical controls sheet 2 foundations

		Historical	Estimate					
Excel estimate CTG iso-phase bus duct foundation (16 ea.)						265.6	120.0 106.4	39.2
Excavation (bathtub)	0.400	0.0	cy	0.0	0.0			
Forms (10 × 10 × 2.5′)	0.300	400	sf	400.0	120.0	120.0		
Rebar	14.0	2.8	ton	2.8	39.2			39.2
Concrete (6′ × 6′ × 2.5′)	1.5	53.3	cy	53.3	80.0		80.0	
Mud mat (10′ × 10′ × 0.33′)	0.500	4.9	cy	4.9	2.5		2.5	
Backfill, compact	0.600	40.0	cy	40.0	24.0		24.0	
Excel estimate STG iso-phase bus duct foundation (16 ea.)						265.6	120.0 106.4	39.2
Excavation (bathtub)	0.400	0.0	cy	0.0	0.0			
Forms (10 × 10 × 2.5′)	0.300	400	sf	400.0	120.0	120.0		
Rebar	14.0	2.8	ton	2.8	39.2			39.2
Concrete (6′ × 6′ × 2.5′)	1.5	53.3	cy	53.3	80.0		80.0	
Mud mat (10′ × 10′ × 0.33′)	0.500	4.9	cy	4.9	2.5		2.5	
Backfill, compact	0.600	40.0	cy	40.0	24.0		24.0	
Excel estimate CTG step-up transformer foundation (16 ea.)						1322.2	453.6 605.4	263.2
Excavation (bathtub)	0.400	0.0	cy	0.0	0.0			
Forms (44 × 64 × 0.25′)	0.300	1512	sf	1512.0	453.6	453.6		
Rebar	14.0	18.8	ton	18.8	263.2			263.2
Concrete (40′ × 60′ × 0.25′)	1.5	355.6	cy	355.6	533.4		533.4	
Mud mat (44′ × 64′ × 0.33′)	0.500	139.1	cy	139.1	69.6		69.6	
Backfill, compact	0.600	4.0	cy	4.0	2.4		2.4	
Excel estimate STG main step-up transformer fen (16 ea.)						1322.2	453.6 605.4	263.2
Excavation (bathtub)	0.400	0.0	cy	0.0	0.0			
Forms (44 × 64 × 0.25′)	0.300	1512	sf	1512.0	453.6	453.6		
Rebar	14.0	18.8	ton	18.8	263.2			263.2
Concrete (40′ × 60′ × 0.25′)	1.5	355.6	cy	355.6	533.4		533.4	
Mud mat (44′ × 64′ × 0.33′)	0.500	139.1	cy	139.1	69.6		69.6	
Backfill, compact	0.600	4.0	cy	4.0	2.4		2.4	
Excel estimate auxiliary transformer (16 ea.)						1303.3	645.0 446.9	211.4
Excavation (bathtub)	0.400	0.0	cy	0.0	0.0			
Forms (34 × 30 × 0.6′)	0.300	2150	sf	2150.0	645.0	645.0		

(Continued)

(Continued)

	Historical		Estimate					
Rebar	14.0	15.1	ton	15.1	211.4			211.4
Concrete (30′ × 26′ × 0.6′)	1.5	277.3	cy	277.3	416.0		416.0	
Mud mat (34′ × 30′ × 0.33′)	0.500	50.4	cy	50.4	25.2		25.2	
Backfill, compact	0.600	9.6	cy	9.6	5.8		5.8	
Excel estimate SUS transformer (5 ea.)					**404.6**	**202.5**	**168.5**	**33.6**
Excavation (bathtub)	0.400	0.0	cy	0.0	0.0			
Forms (16 × 11 × 2.5′)	0.300	675	sf	675.0	202.5	202.5		
Rebar	14.0	2.4	ton	2.4	33.6			33.6
Concrete (12′ × 7′ × 2.5′)	1.5	44.7	cy	44.7	67.1		67.1	
Mud mat (16′ × 11′ × 0.33′)	0.500	10.9	cy	10.9	5.5		5.5	
Backfill, compact	0.600	160.0	cy	160.0	96.0		96.0	

6.4.12 Utility distribution plant electrical controls sheet 3 foundations

	Historical		Estimate					
Excel estimate CTG PDC					**429.5**	**133.5**	**206.4**	**89.6**
Excavation (bathtub)	0.400	0.0	cy	0.0	0.0			
Forms (22 × 67 × 2.5′)	0.300	445	sf	445.0	133.5	133.5		
Rebar	14.0	6.4	ton	6.4	89.6			89.6
Concrete (18′ × 63′ × 2.5′)	1.5	115.5	cy	115.5	173.3		173.3	
Mud mat (22′ × 67′ × 0.33′)	0.500	18.2	cy	18.2	9.1		9.1	
Backfill, compact	0.600	40.0	cy	40.0	24.0		24.0	
Backfill, compact	0.600	160.0	cy	160.0	96.0		96.0	
Excel estimate STG/HRSG PDC					**488.1**	**126.0**	**251.5**	**110.6**
Excavation (bathtub)	0.400	0.0	cy	0.0	0.0			
Forms (47 × 37 × 2.5′)	0.300	420	sf	420.0	126.0	126.0		
Rebar	14.0	7.9	ton	7.9	110.6			110.6
Concrete (43′ × 33′ × 2.5′)	1.5	144.5	cy	144.5	216.8		216.8	
Mud mat (47′ × 37′ × 0.33′)	0.500	21.5	cy	21.5	10.8		10.8	
Backfill, compact	0.600	40.0	cy	40.0	24.0		24.0	
Excel estimate miscellaneous cable tray supports and fdn. (5 ea.)					**247.4**	**192.0**	**45.6**	**9.8**
Excavation (bathtub)	0.400	0.0	cy	0.0	0.0			
Forms (8 × 6 × 1.6′)	0.300	640	sf	640.0	192.0	192.0		
Rebar	14.0	0.7	ton	0.7	9.8			9.8
Concrete (4′ × 2′ × 1.6′)	1.5	13.0	cy	13.0	19.5		19.5	
Mud mat (8′ × 6′ × 0.33′)	0.500	0.6	cy	21.5	10.8		10.8	
Backfill, compact	0.600	25.6	cy	25.6	15.4		15.4	
Excel estimate transmission towers (6 ea.)					**412.2**	**210.6**	**151.2**	**50.4**
Excavation (bathtub)	0.400	0.0	cy	0.0	0.0			
Forms (13 × 10 × 4.5′)	0.300	702	sf	702.0	210.6	210.6		
Rebar	14.0	3.6	ton	3.6	50.4			50.4
Concrete (9′ × 6′ × 4.5′)	1.5	64.8	cy	64.8	97.2		97.2	
Mud mat (13′ × 10′ × 0.33′)	0.500	1.6	cy	21.5	10.8		10.8	
Backfill, compact	0.600	72.0	cy	72.0	43.2		43.2	
					406.4	**239.4**	**123.6**	**43.4**

(Continued)

(Continued)

		Historical	Estimate					
Excel estimate takeoff towers (7 ea.)								
Excavation (bathtub)	0.400	0.0	cy	0.0	0.0			
Forms (19 × 8 × 3')	0.300	798	sf	798.0	239.4	239.4		
Rebar	14.0	3.1	ton	3.1	43.4			43.4
Concrete (15' × 4' × 3')	1.5	56.0	cy	56.0	84.0		84.0	
Mud mat (19' × 8" × 0.33')	0.500	1.9	cy	21.5	10.8		10.8	
Backfill, compact	0.600	48.0	cy	48.0	28.8		28.8	
Excel estimate disconnect switch					**131.2**	**51.6**	**62.8**	**16.8**
Excavation (bathtub)	0.400	0.0	cy	0.0	0.0			
Forms (14 × 29 × 2')	0.300	172	sf	172.0	51.6	51.6		
Rebar	14.0	1.2	ton	1.2	16.8			16.8
Concrete (10' × 25' × 2')	1.5	21.9	cy	21.9	32.9		32.9	
Mud mat (14' × 29" × 0.33')	0.500	5.0	cy	21.5	10.8		10.8	
Backfill, compact	0.600	32.0	cy	32.0	19.2		19.2	

6.4.13 Utility distribution plant electrical controls sheet 4

		Historical	Estimate					
Excel estimate circuit breaker					**143.5**	**63.0**	**67.9**	**12.6**
Excavation (bathtub)	0.400	0.0	cy	0.0	0.0			
Forms (19 × 11 × 3.5')	0.300	210	sf	210.0	63.0	63.0		
Rebar	14.0	0.9	ton	0.9	12.6			12.6
Concrete (15' × 7' × 3.5')	1.5	15.7	cy	15.7	23.6		23.6	
Mud mat (19' × 11' × 0.33')	0.500	2.6	cy	21.5	10.8		10.8	
Backfill, compact	0.600	56.0	cy	56.0	33.6		33.6	
Excel estimate pole supports (14 ea.)					**136.2**	**75.6**	**46.6**	**14.0**
Excavation (bathtub)	0.400	0.0	cy	0.0	0.0			
Forms (9 × 9 × 1')	0.300	252	sf	252.0	75.6	75.6		
Rebar	14.0	1	ton	1.0	14.0			14.0
Concrete (5' × 5' × 1')	1.5	17.5	cy	17.5	26.3		26.3	
Mud mat (9' × 9" × 0.33')	0.500	1.0	cy	21.5	10.8		10.8	
Backfill, compact	0.600	16.0	cy	16.0	9.6		9.6	
Excel estimate miscellaneous small foundations (10 ea.)					**298.8**	**216.0**	**64.6**	**18.2**
Excavation (bathtub)	0.400	0.0	cy	0.0	0.0			
Forms (9 × 9 × 2')	0.300	720	sf	720.0	216.0	216.0		
Rebar	14.0	1.3	ton	1.3	18.2			18.2
Concrete (5' × 5' × 2')	1.5	23.1	cy	23.1	34.7		34.7	
Mud mat (9' × 9" × 0.33')	0.500	1.0	cy	21.5	10.8		10.8	
Backfill, compact	0.600	32.0	cy	32.0	19.2		19.2	
Excel estimate electrical manholes (7 ea.)					**1679.2**	**861.8**	**562.6**	**254.8**
Excavation (bathtub)	0.400	0.0	cy	0.0	0.0			
Forms (19 × 17 × 4.5')	0.300	2872.8	sf	2872.8	861.8	861.8		
Rebar	14.0	18.2	ton	18.2	254.8			254.8
Concrete (9' × 6' × 4.5')	1.5	331.4	cy	331.4	497.1		497.1	
Mud mat (19' × 17" × 0.33')	0.500	4.0	cy	21.5	10.8		10.8	
Backfill, compact	0.600	91.2	cy	91.2	54.7		54.7	

(Continued)

(Continued)

	Historical			Estimate				
Excel estimate transmission towers (6 ea.)					412.2	210.6	151.2	50.4
Excavation (bathtub)	0.400	0.0	cy	0.0	0.0			
Forms (13 × 10 × 4.5′)	0.300	702	sf	702.0	210.6	210.6		
Rebar	14.0	3.6	ton	3.6	50.4			50.4
Concrete (9′ × 6′ × 4.5′)	1.5	64.8	cy	64.8	97.2		97.2	
Mud mat (13′ × 10″ × 0.33′)	0.500	1.6	cy	21.5	10.8		10.8	
Backfill, compact	0.600	72.0	cy	72.0	43.2		43.2	
Excel estimate transmission towers (6 ea.)					846.1	126.0	360.3	359.8
Excavation (bathtub)	0.400	0.0	cy	0.0	0.0			
Forms (5 × 7.8 × 7′)	0.300	420	sf	420.0	126.0	126.0		
Rebar	14.0	25.7	ton	25.7	359.8			359.8
Concrete (16′ × 7.8′ × 7′)	1.5	233.0	cy	233.0	349.5		349.5	
Mud mat (5′ × 7.8″ × 0.33′)	0.500	0.5	cy	21.5	10.8		10.8	
Backfill, compact	0.600	0.0	cy	0.0	0.0		0.0	
Excel estimate trenches, curbs, equip pads, piers, etc.					3021.8	0.0	1999.8	1022.0
Excavation (bathtub)	0.400	0.0	cy	0.0	0.0			
Forms (5 × 7.8 × 7′)	0.300	0	sf	0.0	0.0	0.0		
Rebar	14.0	73	ton	73.0	1022.0			1022.0
Concrete (16′ × 7.8′ × 7′)	1.5	1326.0	cy	1326.0	1989.0		1989.0	
Mud mat (5′ × 7.8″ × 0.33′)	0.500	0.0	cy	21.5	10.8		10.8	
Backfill, compact	0.600	0.0	cy	0.0	0.0		0.0	

6.5 Summary foundation man-hours

Direct craft man-hours	Estimate			
Scope of work	MH	Carpenter	Labor	IW
Foundations	**53,530**	**13,081**	**29,452**	**10,997**
Boiler foundations	7347	1138	4550	1659
CTG and CTG-related foundations	5528	1025	3330	1173
STG and STG-related foundations	15,865	2939	9471	3455
Plant mechanical system−related foundations	7161	2046	3962	1152
Utility distribution plant electrical controls foundations	17,628	5932	8138	3557

6.6 Mechanical equipment (CTG, STG, heat recovery steam generator) includes vendor piping

6.7 HRSG triple pressure; three wide-installation man-hours

HRSG triple pressure; three wide-installation man-hours	59,600
Excel spreadsheet summary HRSG vendor piping (PF)	35,341
Excel spreadsheet summary HRSG equipment (BM)	24,258

HRSG triple pressure; three wide-installation equipment man-hours	24,258
HRSG—module casing	6528
HRSG—ductwork SCR and inlet	3606
HRSG—erect gas baffles	2578
HRSG—modules	1111
HRSG—pressure vessels	420
HRSG—install duct burner assemblies and elements	1160
HRSG—steel (ladders, platforms, and grating)	4300
HRSG—skids	240
HRSG—stack and breeching	2898
HRSG—SCR and CO_2 internals	1417

6.8 Estimate heat recovery steam generator triple pressure; three wide

		Historical		Estimate	
Description	MH	Qty	Unit	Qty	BM
	MH_a	n_i		n_i	$n_i\,MH_a$
Estimate—heat recovery steam generator					**24,258**
6.8.1 HRSG—module casing					**6528**
Slide plate assemblies					
Shim and set slide/base plates (provide shims)	1.00	45.0	EA	45.0	45.0
Weld shear blocks	0.50	180.0	EA	180.0	90.0
Grout base plates	4.00	45.0	EA	45.0	180.0
Anchor bolts	0.65	180.0	EA	180.0	117.0
Erect casing panels and assemblies					
Casing assembly—columns, floor beams	20.00	35.0	EA	35.0	700.0
Roof beam corner angle	4.00	24.0	EA	24.0	96.0
Bellows	2.60	40.0	EA	40.0	104.0
Install roof/floor beam assemblies					
Roof beams assembly	25.00	24.0	EA	24.0	600.0
Frame connections					
Moment welds					
Moment weld-col 13/16″ single groove/3/16″ fillet	12.00	128.0	EA	128.0	1536.0
Boltup 7/8″ diameter × 2″	1.00	96.0	EA	96.0	96.0
Fillet weld at web 1/4″	2.00	64.0	EA	64.0	128.0
Remove lift lug	0.50	32.0	EA	32.0	16.0
Remove angle clip	0.50	32.0	EA	32.0	16.0

(Continued)

(Continued)

Description	MH	Historical Qty	Unit	Estimate Qty	BM
	MH_a	n_i		n_i	$n_i\,MH_a$
Field seam liners and insulation					
Install ceramic fiber blanket	0.04	4663.4	SF	4663.4	186.5
Install ogee clips, washers, bolt, and weld washer	0.25	460.0	EA	460.0	115.0
Install mesh SS screen wire	0.04	4663.4	SF	4663.4	186.5
Install insulation and liner plates	1.00	460.0	SF	460.0	460.0
Weld casing seams					
Casing assembly w/cols	0.35	1537.0	LF	1537.0	538.0
Casing assembly w/floor beams	0.35	897.0	LF	897.0	314.0
Roof casing w/tube bundle assembly	0.35	1255.8	LF	1255.8	439.5
Casing panel	0.35	973.6	LF	973.6	340.8
Fit up	0.05	4663.4	LF	4663.4	223.8

6.8.2 HRSG—ductwork SCR and inlet					3606
Slide plate assemblies (SCR and inlet)					
Shim and set slide/base plates (provide shims)	1.00	12.0	EA	12.0	12.0
Weld sheer blocks	0.50	48.0	EA	48.0	24.0
Grout base plates	4.00	12.0	EA	12.0	48.0
Anchor bolts	0.65	48.0	EA	48.0	31.2
Erect casing panels and assemblies					
Casing assembly, flange assembly, and grid assembly	10.00	24.0	EA	24.0	240.0
Floor panel	10.00	8.0	EA	8.0	80.0
Casing assembly w/roof beam	20.00	6.0	EA	6.0	120.0
Install roof/floor beam assemblies					
SCR					
Roof beams assembly	25.00	6.0	EA	6.0	150.0
Moment welds					
Frame connection: inlet duct "A," "B," "C," and "D"					
Moment weld-col 13/16" single groove/3/16" fillet	12.00	48.0	EA	48.0	576.0
Moment weld-col 9/16" single groove/3/16" fillet	10.00	32.0	EA	32.0	320.0
Moment weld-col 5/8" single groove/3/16" fillet	8.00	16.0	EA	16.0	128.0
Boltup 7/8" diameter × 2"	1.00	48.0	EA	48.0	48.0
Fillet weld at web 1/4"	2.00	31.5	EA	31.5	63.0
Remove lift lug	0.50	32.0	EA	32.0	16.0
Remove angle clip	0.50	48.0	EA	48.0	24.0

(*Continued*)

(Continued)

6.8.2 HRSG—ductwork SCR and inlet					3606
Field seam liners and insulation					
Inlet duct					
Field joint at wall					
Install insulation and liner plates	1.00	150.0	EA	150.0	150.0
Install ogee clips, washers, bolt, and weld washer	0.25	150.0	EA	150.0	37.5
Field joint at floor and roof					
Install insulation and liner plates	1.00	120.0	EA	120.0	120.0
Install ogee clips, washers, bolt, and weld washer	0.25	120.0	EA	120.0	30.0
SCR/CO					
Insulate and install lower liner plates at field seams					
Plates	1.00	400.0	EA	400.0	400.0
Insulation	0.04	4800.0	SF	4800.0	192.0
Weld casing seams					
Spool duct					
Casing panel	0.35	307.4	LF	307.4	107.6
Casing panel	0.35	175.4	LF	175.4	61.4
Roof casing w/tube bundle assembly	0.35	175.4	LF	175.4	61.4
SCR					
Casing assembly w/cols	0.35	307.4	LF	307.4	107.6
Casing assembly w/floor beams	0.35	165.4	LF	165.4	57.9
Roof casing w/tube bundle assembly	0.35	165.4	LF	165.4	57.9
Inlet duct					
Field joint at wall					
Seal weld secondary casing to column	0.35	440.0	LF	440.0	154.0
Field joint at floor and roof					
Seal weld secondary casing to column	0.35	270.0	LF	270.0	94.5
Cover plate-floor/roof	0.35	270.0	LF	270.0	94.5

6.8.3 HRSG—erect gas baffles, pressure vessels, modules, and duct burner					5269
Install side wall gas baffles					
Baffle plates	0.75	880.0	EA	880.0	660.0
Fillet weld	0.35	1320.0	LF	1320.0	462.0
Install upper splice gas baffles					
Splice plates	0.75	647.0	EA	647.0	485.3
U-bolt connections	0.35	647.0	EA	647.0	226.5
Install lower gas baffles					
Baffle plates	0.75	307.0	EA	307.0	230.3
U-bolt connections	0.35	307.0	EA	307.0	107.5
Header/shipping restraints					
Angle 4 × 4 × 3/4 × 6'-1-5/8"	1.00	78.0	EA	78.0	78.0
Tube bundle gas baffles					
Baffle plates	0.75	299.0	EA	299.0	224.3

(Continued)

(Continued)

6.8.3 HRSG—erect gas baffles, pressure vessels, modules, and duct burner					5269
Fillet weld	0.35	299.0	LF	299.0	104.7
Estimate—modules					
HRSG—modules					
Reheater 3 HP superheated 2/reheater 2	2.20	56.0	ton	56.0	123.2
HP superheated 1/reheater 1	1.80	65.5	ton	65.5	117.9
HP evaporator/IP superheated 2	1.40	165.5	ton	165.5	231.7
HP economizer 2/IP superheated 1	1.50	147.0	ton	147.0	220.5
HP economizer 1/IP economizer 1/LP evaporator	1.50	138.5	ton	138.5	207.8
Feedwater heater 2	2.20	55.0	ton	55.0	121.0
Feedwater heater 1	2.50	35.5	ton	35.5	88.8
Estimate—pressure vessels					
HRSG—pressure vessels					
HP steam drum					
Set saddle base plates	24.00	1.0	EA	1.0	24.0
Set drum	1.40	158.0	ton	158.0	220.4
IP steam drum					
Set saddle base plates	24.00	1.0	EA	1.0	24.0
Set drum	2.50	9.0	ton	9.0	22.5
LP steam drum					
Set saddle base plates	24.00	1.0	EA	1.0	24.0
Set drum	2.20	30.0	ton	30.0	66.0
Flash separator					
Set saddle base plates	24.00	1.0	EA	1.0	24.0
Set drum	2.50	6.0	ton	6.0	15.0
Estimate—duct burner					
HRSG—install duct burner assemblies and elements					
Duct burner assembly	20.00	18.0	EA	18.0	360.0
Scanner, hose, and retainer assembly	10.00	55.0	EA	55.0	550.0
Element support	5.00	50.0	EA	50.0	250.0

6.8.4 HRSG—steel (ladders, platforms, and grating), skids, SCR, and CO					5957
Assemble and install all of the structural steel and platforms					
Top of HRSG L′ × W′	0.40	4140.0	SF	4140.0	1656.0
HP, IP, LP drums L′ × W′	0.40	540.0	SF	540.0	216.0
Top of inlet duct L′ × W′	0.40	1080.0	SF	1080.0	432.0
Right/left side X EA times (L′ × W′)	0.40	518.4	SF	518.4	207.4
Stair tower X EA times (L′ × W′)	0.40	384.0	SF	384.0	153.6
Misc X EA times (L′ × W′)	0.40	216.0	SF	216.0	86.4
Handrail LF	0.32	714.0	LF	714.0	228.5
Ladders LF	1.00	360.0	LF	360.0	360.0
Stair tower structural	32.00	30.0	ton	30.0	960.0

(Continued)

(Continued)

6.8.4 HRSG—steel (ladders, platforms, and grating), skids, SCR, and CO					5957
Estimate—skids					
HRSG—skids					
Level and set the AFCU skid	80.00	1.0	EA	1.0	80.0
Level and set the piping module	80.00	1.0	EA	1.0	80.0
Level and set the blower skid	80.00	1.0	EA	1.0	80.0
Estimate—SCR and CO					
HRSG—SCR and CO_2 internals					
Install internal structure					
Remove shipping braces	1.00	100.0	EA	100.0	100.0
Set support structure and cut/weld and grind lifting lugs	30.00	2.0	EA	2.0	60.0
Set left, middle right internal structure	30.00	6.0	EA	6.0	180.0
Weld structure bases to each other	0.35	560.0	LF	560.0	196.0
Insert and weld top and side pins-4"-1/4" fillet weld	0.35	18.0	EA	18.0	6.3
Catalyst anchor installation					
Set anchor channel and weld to structure	0.35	216.0	LF	216.0	75.6
Install T-shaped anchors	0.75	216.0	EA	216.0	162.0
Make up T-shaped-washer/nut w/jam nut	0.35	216.0	EA	216.0	75.6
Seal plate installation					
Install side/top/bottom slide plates	2.00	62.0	EA	62.0	124.0
Weld seal plates	0.35	1060.0	LF	1060.0	371.0
Trim seal plates	0.25	265.0	LF	265.0	66.3

6.8.5 HRSG—stack and breeching (XX'-X" OD × XXX'-X" high)					2898
Erect lower stack					
Seam up shell cans for lower section					
Set shell cans (field fabrication)	4.00	16.0	EA	16.0	64.0
Weld vertical (shell can to shell can w/stiffener splices)	0.75	224.0	LF	224.0	168.2
Stack shell cans and weld to make up lower sections					
Set shell cans	4.00	16.0	EA	16.0	64.0
Weld horizontal	0.75	450.0	LF	450.0	337.5
Shoot shim packs, erect lower stack, and grout base ring					
Shims and bolt to foundation	2.40	32.0	EA	32.0	76.8
Set lower stack section	60.00	1.0	EA	1.0	60.0
Load and haul to erection site	4.00	28.0	EA	28.0	112.0
Erect damper					
Seam up (2) damper halves					
Set up damper halves (field fabrication)	10.00	2.0	EA	2.0	20.0
Weld damper halves	0.80	20.0	LF	20.0	16.0

(Continued)

(Continued)

6.8.5 HRSG—stack and breeching (XX'-X" OD × XXX'-X" high)					2898
Boltup	0.40	100.0	EA	100.0	40.0
Load and haul dampers to erection site	16.00	2.0	EA	2.0	32.0
Erect upper stack					
Seam up shell cans for upper section					
Set shell cans (field fabrication)	4.00	14.0	EA	14.0	56.0
Weld vertical (shell can to shell can w/stiffener splices)	0.75	220.0	LF	220.0	165.2
Stack shell cans and weld to make up upper sections					
Set shell cans	4.00	16.0	EA	16.0	64.0
Weld horizontal	0.50	460.0	LF	460.0	230.0
Erect upper stack sections					
Set upper sections	60.00	1.0	EA	1.0	60.0
Weld horizontal	0.75	120.0	LF	120.0	90.0
Load and haul to erection site	4.00	20.0	EA	20.0	80.0
Attach structural steel and platforms					
360-Degree platform	0.15	360.0	EA	360.0	54.0
Ladder	0.35	132.0	LF	132.0	46.2
360-Degree platform	0.15	360.0	SF	360.0	54.0
Bridge	0.15	240.0	SF	240.0	36.0
Fit and weld breech sections to stack					
Set panels	60.00	2.0	EA	2.0	120.0
Weld	0.75	330.0	LF	330.0	247.8
Install expansion joint					
Expansion joint	60.00	1.0	EA	1.0	60.0
Backup bars	1.00	24.0	EA	24.0	24.0
Boltup	0.25	1120.0	EA	1120.0	280.0
Expanded metal	1.20	200.0	EA	200.0	240.0

6.9 Heat recovery steam generator large bore vendor piping

Excel spreadsheet summary HRSG vendor piping	14,145	lf	35,341	2.50
Excel summary HRSG LB and SB vendor piping	7343	lf	20,720	2.82

	Qty		PF	
	n_i	Unit	$n_i\,MH_a$	MH/LF
HRSG—large bore code piping	**4238**	**lf**	**15,529**	**3.66**
HP piping and supports—ASME Section 1	519	lf	3182	3.10
IP piping and supports—ASME Section 1	136	lf	403	2.96
LP piping and supports—ASME Section 1	134	lf	300	2.24
RP piping and supports—ASME Section 1	177	lf	1548	8.75
HP piping and Supports—ASME B31.1	504	lf	3031	6.01
IP piping and supports—ASME B31.1	646	lf	1022	1.58
LP piping and supports—ASME B31.1	944	lf	2051	2.17
RP piping and supports—ASME B31.1	438	lf	2191	5.00

(Continued)

(Continued)

Excel spreadsheet summary HRSG vendor piping					14,145	lf	35,341	2.50
Excel summary HRSG LB and SB vendor piping					7343	lf	20,720	2.82

	Qty	PF
	n_i	Unit n_i MH$_a$ MH/LF

	Qty	Unit	n_i MH$_a$	MH/LF
	n_i			
Silencer drain piping and supports	740	lf	1800	2.43
HRSG—large bore code piping	**4238.0**	**lf**		**15,529.4**

HP piping and supports—ASME Section 1	**519**	**lf**		**3182**

Line no.	Material	Size	Sch/Thk	MH MH_a	Qty n_i	Unit	Qty n_i	PF $n_i MH_a$
HP-03 econ 1 to HP econ 2	SA-106-B	8	140/.906	2.2	25	lf	25	54.9
HP-03 econ 1 to HP econ 2	SA-106-B	6	160/.906	4.1	42	lf	42	173.2
HP-04 econ 2 to HP steam drum	SA-106-B	8	140/.812	3.1	96	lf	96	295.1
HP-04 econ 2 to HP steam drum	SA-106-C	6	160/.906	31.1	1	lf	1	31.1
HP-09 steam drum to HP SH 1	SA-106-B	6	160/.906	3.8	24	lf	24	90.7
HP-10 steam drum to HP SH 1	SA-106-B	6	160/.906	3.4	23	lf	23	79.3
HP-11 steam drum to HP SH 1	SA-106-B	6	160/.906	3.9	23	lf	23	89.2
HP-12 steam drum to HP SH 1	SA-106-B	6	160/.906	3.4	23	lf	23	79.3
HP-13 steam drum to HP SH 1	SA-106-B	6	160/.906	3.9	23	lf	23	89.2
HP-14 steam drum to HP SH 1	SA-106-B	6	160/.906	3.4	23	lf	23	79.3
HP-15 SHTR 1 to HP SHTR 2	SA-335-P91	10	1.25	9.4	68	lf	68	638.5
HP-15 SHTR 1 to HP SHTR 2	SA-335-P91	8	1	24.9	3	lf	3	74.6
HP-15 SHTR 1 to HP SHTR 2	SA-335-P91	3		0.6	1	lf	1	0.6
HP-16 SHTR 1 to HP SHTR 2	SA-335-P91	10	1.25	9.1	68	lf	68	618.5
HP-16 SHTR 1 to HP SHTR 2	SA-335-P91	8	1	24.9	3	lf	3	74.6
HP-16 SHTR 1 to HP SHTR 2	SA-335-P91	3		0.6	1	lf	1	0.6
HP-17 SHTR 1 to HP SHTR 2	SA-335-P91	10	1.25	9.4	68	lf	68	638.5
HP-17 SHTR 1 to HP SHTR 2	SA-335-P91	8	1	24.9	3	lf	3	74.6
HP-17 SHTR 1 to HP SHTR 2	SA-335-P91	3		0.6	1	lf	1	0.6
IP piping and supports—ASME Section 1					**136.0**	**lf**		**403.0**
IP-03 econ to IP steam drum	SA-106-B	4	40	2.48	53	lf	53	131.2
IP-03 econ to IP steam drum	SA-106-B	3	40	4.54	6	lf	6	27.3
IP-06 drum to IP superheater 1	SA-106-B	10	40	2.37	42	lf	42	99.6
IP-06 drum to IP superheater 1	SA-106-B	6	40	13.37	5	lf	5	66.9
IP-09 SH 1 to IP SH 2	SA-106-B	10	40	2.61	30	lf	30	78.2
LP piping and supports—ASME Section 1					**134.0**	**lf**		**300.2**
LP-10 steam drum to LP SHTR 1	SA-106-B	14	std	2.22	26	lf	26	57.7
LP-10 steam drum to LP SHTR 1	SA-106-B	10	40	2.32	34	lf	34	78.8
LP-10 steam drum to LP SHTR 1	SA-106-B	8	40	3.11	12	lf	12	37.3
LP-14 SHTR 1 to LP SHTR 2	SA-106-B	10	40	2.01	32	lf	32	64.4
LP-15 SHTR 1 to LP SHTR 2	SA-106-B	10	40	2.07	30	lf	30	62.0
RP piping and supports—ASME Section 1					**177.0**	**lf**		**1548.0**
RP-03 reheater 1 to reheater 2	SA-335-P91	24	80/1.218	5.44	137	lf	137	745.5
RP-03 reheater 1 to reheater 2	SA-335-P91	24	80/.843	39.87	20	lf	20	797.4
RP-03 reheater 1 to reheater 2		3		0.26	20	lf	20	5.1

6.9.1 Piping and supports—ASME B31.1 sheet 1

HP piping and supports—ASME B31.1					504.0	lf		3031.4
HP-01 feedwater inlet	SA-106-C	10	160/1.125	3.09	47	lf	47	145.0
HP-01 feedwater inlet	SA-106-C	6	160/.718	9.00	3	lf	3	27.0
HP-01 feedwater inlet	SA-106-C	10	1.5	5.09	57	lf	57	290.0
HP-02-steam outlet	SA-335-P91	14	1.75	11.91	44	lf	44	523.9
HP-02-steam outlet	SA-335-P91	10	1.25	8.65	26	lf	26	224.9
HP-18 SHP-PSV-104 to SHP-S1-002 (stm out)		6		0.39	20	lf	20	7.8
HP-19 start-up vent piping	SA-335-P91	8	120/.718	4.58	22	lf	22	100.7
HP-20 sparging steam header	SA-106-B	3	160/.437	1.69	42	lf	42	71.0
HP-21 SHP-PSV-112A to SHP-S1-001 (Drum)		6		1.92	20	lf	20	38.4
HP-22 SHP-PSV-112B to SHP-S1-001 (Drum)		6		1.92	20	lf	20	38.4
HP-23 upper drain piping coll header	SA-106-B	8	40	10.56	20	lf	20	211.2
HP-23 upper drain piping coll header	SA-106-B	3	40	0.46	66	lf	66	30.2
HP-24 steam drum warming connection	SA-106-B	4	160	3.12	16	lf	16	49.9
HP-25 SHP-PSV-104 stack piping	SA-106-B	14	40	2.72	29	lf	29	78.8
HP-25 SHP-PSV-104 stack piping	SA-106-B	16	40	21.12	20	lf	20	422.4
HP-26 SHP-PSV-112A stack piping	SA-106-B	18	40	23.76	20	lf	20	475.2
HP-26 SHP-PSV-112A stack piping	SA-106-B	16	40	6.05	12	lf	12	72.6
HP-27 SHP-PSV-112B stack piping	SA-106-B	18	20	4.32		lf	20	86.4
HP-27 SHP-PSV-112B stack piping	SA-106-B	16	12	6.88	20	lf	20	137.6
IP piping and supports—ASME B31.1					646.0	lf		1021.7
IP-01 feedwater inlet	SA-106-B	4	40	1.06	136	lf	136	143.9
IP-02-steam outlet	SA-106-B	10	40	1.88	84	lf	84	157.7
IP-02-steam outlet	SA-106-B	8	40	11.14	4	lf	4	44.6
IP-10 pegging steam	SA-106-B	16	std	18.88	7	lf	7	132.2
IP-10 pegging steam	SA-106-B	6	40	2.46	18	lf	18	44.3
IP-11 SIP-PSV-109 to SIP-SI-002 (stm out)		4		0.90	20	lf	20	18.0
IP-12 start-up to SIP-SI-005	SA-106-B	6	40	3.51	19	lf	19	66.7
IP-13 SIP-PSV-106 to SIP-SI-001 (drum)		6		1.35	20	lf	20	27.0
IP-15 sparging steam header	Sa-106-B	3	40	0.77	38	lf	38	29.2
IP-16 intermittent blow off	SA-106-B	2.5	80	0.85	84	lf	84	71.6
IP-17 SIP-PSV-109 stack piping	SA-106-B	10	40	7.30	20	lf	20	146.0
IP-17 SIP-PSV-109 stack piping	SA-106-B	8	40	1.02	20	lf	20	20.4
IP-18 SIP-PSV-106 stack piping	SA-106-B	12	40	0.96	20	lf	20	19.2
IP-18 SIP-PSV-106 stack piping	SA-106-B	10	40	1.39	16	lf	16	22.3
IP-19 flash tank 1 to IP drum	SA-106-B	2.5	80	0.46	18	lf	18	8.4
IP-20 F.S. 1 to blowdown tank	SA-106-B	6	80	0.48	20	lf	20	9.6
IP-21 BFW-PSV 200 outlet piping	SA-106-B	2.5	160	0.59	102	lf	102	60.7

6.9.2 Piping and supports—ASME B31.1 sheet 2

LP piping and supports—ASME B31.1					944.0	lf		2051.1
LP-01 feedwater inlet	SA-106-B	8	40	1.83	109	lf	109	199.1
LP-01 feedwater inlet	SA-106-B	6	40	2.34	20	lf	20	46.8
LP-02 steam outlet	SA-106-B	14	std	3.41	65	lf	65	221.4

(*Continued*)

(Continued)

LP piping and supports—ASME B31.1						944.0	lf		2051.1
LP-02 steam outlet	SA-106-B	10	40		5.76	23	lf	23	132.4
LP-02 steam outlet	SA-106-B	8	std		10.24	20	lf	20	204.8
LP-02 steam outlet		3			0.12	20	lf	20	2.4
LP-03 LP FW heater bypass	SA-106-B	8	40		16.77	3	lf	3	50.3
LP-04 FW heater 1 to FW heater 2	SA-106-B	6	40		4.90	5	lf	5	24.5
LP-05 FW heater 1 to FW heater 2	SA-106-B	6	40		4.90	5	lf	5	24.5
LP-06 FW heater 1 to FW heater 2	SA-106-B	6	40		4.90	5	lf	5	24.5
LP-07 FW heater 2 to LP steam drum	SA-106-B	8	40		4.64	45	lf	45	208.8
LP-07 FW heater 2 to LP steam drum	SA-106-B	6	40		7.38	6	lf	6	44.3
LP-16 boiler feed pump recirc.	SA-106-B	6	40		1.11	49	lf	49	54.5
LP-16 boiler feed pump recirc.	SA-106-B	6	1		2.10	6	lf	6	12.6
LP-17 boiler feed pump recirc.	SA-106-B	6	40		1.20	43	lf	43	51.6
LP-17 boiler feed pump recirc.	SA-106-B	6	1		2.10	6	lf	6	12.6
LP-20 RAC to LP steam drum	SA-106-B	10	40		1.58	203	lf	203	320.4
LP-21 SLP-PSV-102 to SLP-SI-002 (stm out)		6			1.62	20	lf	20	32.4
LP-22 start-up vent to SLP-SI-003	SA-106-B	8	40		3.71	19	lf	19	70.4
LP-23 SLP-PSV-104A to SLP-SI-001 (drum)		6			1.62	20	lf	20	32.4
LP-24 SLP-PSV-104B to SLP-SI-001 (drum)		6			1.62	20	lf	20	32.4
LP-26 sparging steam header	SA-106-B	3	40		1.04	38	lf	38	39.5
LP-27 intermittent blow off	SA-106-B	2.5	80		0.81	91	lf	91	73.7
LP-28 SLP-PSV-102 stack piping	SA-106-B	10	40		0.80	20	lf	20	16.0
LP-28 SLP-PSV-102 stack piping	SA-106-B	8	40		1.50	20	lf	20	30.0
LP-29 SLP-PSV-104A stack piping	SA-106-B	10	40		0.80	20	lf	20	16.0
LP-29 SLP-PSV-104A stack piping	SA-106-B	8	40		2.71	12	lf	12	32.5
LP-30 SLP-PSV-104B stack piping	SA-106-B	10	40		0.80	20	lf	20	16.0
LP-30 SLP-PSV-104B stack piping	SA-106-B	8	40		2.20	11	lf	11	24.2
RP piping and supports—ASME B31.1						438.0	lf		2191.4
RP-01 reheater 1 inlet	SA-106-B	24	60/1.218		6.08	60	lf	60	364.6
RP-01 reheater 1 inlet	SA-106-B	16	60/.843		26.40	14	lf	14	369.6
RP-01 reheater 1 inlet		6			0.48	20	lf	20	9.6
RP-02 reheater 3 outlet	SA-335-P91	24	60/1.218		6.84	50	lf	50	342.0
RP-02 reheater 3 outlet	SA-335-P91	16	60/.843		14.43	24	lf	24	346.3
RP-02 reheater 3 outlet	SA-335-P91	10	40		1.50	20	lf	20	30.0
RP-02 reheater 3 outlet		4			0.16	20	lf	20	3.2
RP-09 SIP-PSV-108A to SIP-SI-003 (CLD RH)		6			1.62	20	lf	20	32.4
RP-10 SIP-PSV-108B to SIP-SI-003 (CLD RH)		6			1.62	20	lf	20	32.4
RP-11 SIP-PSV-105 to SIP-SI-004 (RH out)		6			1.62	20	lf	20	32.4
RP-12 start-up vent	SA-335-P91	10	40		6.90	27	lf	27	186.4
RP-13 SIP-PSV-108A stack piping	SA-106-B	18	40		2.70	20	lf	20	54.0
RP-13 SIP-PSV-108A stack piping	SA-106-B	16	40		5.01	19	lf	19	95.2
RP-13 SIP-PSV-108B stack piping	SA-106-B	20	40		3.00	20	lf	20	60.0
RP-13 SIP-PSV-108B stack piping	SA-106-B	18	40		6.19	16	lf	16	99.0
RP-15 SIP-PSV-105 stack piping	SA-106-B	14	40		2.10	20	lf	20	42.0
RP-15 SIP-PSV-105 stack piping	SA-106-B	12	40		3.66	20	lf	20	73.2
RP-03 cooling air piping	SA-106-B	3	40		0.68	28	lf	28	19.2

6.9.3 Piping and supports—ASME B31.1 sheet 3

Silencer drain piping and supports						740.0	lf		1800.3
14"-RP-03 PRVS-X703B to SIL-X704	SA-106-B	14	40	3.04	40		lf	40	121.7
10"-RP-05 PRVS-X703A to SIL-X704	SA-106-B	14	40	3.04	40		lf	40	121.7
10"-RP-10 PRVS-X704 to SIL-X705	SA-106-B	10	40	2.55	40		lf	40	101.9
12"-RP-08 MOV-X211/212 SIL-X706	SA-106-B	12	40	4.49	40		lf	40	179.5
10"-HP-13 PRVS-X802 to SIL-X803	SA-106-B	10	40	2.17	40		lf	40	86.9
12"-HP-17 PRVS-X801A to SIL-X801	SA-106-B	12	40	2.61	40		lf	40	104.3
8"-LP-13 PRVS-X602 to SIL-X603	SA-106-B	8	40	1.74	40		lf	40	69.5
6"-IP-10 PRVS-X702 to SIL-X703	SA-106-B	6	40	1.30	40		lf	40	52.1
4"-IP-08 MOV-X205/206 to SIL-X702	SA-106-B	4	40	1.50	40		lf	40	59.8
8"-IP-12 PRVS-701 to SIL-X701	SA-106-B	8	40	1.74	40		lf	40	69.5
12"-LP-11 MOV-X106/107 to SIL-X602	SA-106-B	12	40	4.49	40		lf	40	179.5
8"-LP-15 PRVS-X601A to SIL-X601	SA-106-B	8		1.74	40		lf	40	69.5
8"-LP-17 PRVS-X601B to SIL-X601	SA-106-B	8		1.74	40		lf	40	69.5
12"-HP-15 PRVS-X801B to SIL-X801	SA-106-B	12	40	2.61	40		lf	40	104.3
6"-IP-11 MOV-X310/311 to SIL-X803	SA-106-B	6	40	2.40	40		lf	40	96.0
PSV Drip Pan Assembly	SA-106-B	6	40	2.25	140		lf	140	314.5

6.10 Heat recovery steam generator small bore vendor piping

6.10.1 Excel spreadsheet summary heat recovery steam generator vendor piping

		Qty		PF
	n_i	Unit	$n_i MH_a$	MH/LF
HRSG—small bore code piping	**3105**	**lf**	**5191**	**1.67**
HP piping and supports	760	lf	1125	1.48
IP piping and supports	700	lf	957	1.37
LP piping and supports	420	lf	580	1.38
RP piping and supports	385	lf	604	1.57
Blowdown piping	680	lf	880	1.29
Instrumentation	160	lf	1044	6.53

6.10.2 Heat recovery steam generator—small bore code piping

					2945	lf		4147
HP piping and supports					760	lf		1125
				MH	Qty	Unit	Qty	PF
Line no.	Material	Size	Sch/Thk	MH$_a$	n$_i$		n$_i$	n$_i$ MH$_a$
HP-51 superheater 2 drain	SA-106-B	2	80	1.87	80	lf	80	149.3
HP-50 superheater 1 drain	SA-106-B	2	80	1.69	80	lf	80	135.5
HP economizer 1 drain	SA-106-B	2	80	1.58	140	lf	140	221.6
HP economizer 2 drains	SA-106-B	2	80	1.34	220	lf	220	295.5
HP-18-A sparging header	SA-106-B	2	80	2.57	20	lf	20	51.3
HP-23-D manifold drain	SA-106-B	2	80	2.27	20	lf	20	45.4
HP-52 upper blowdown piping	SA-106-B	2	80	1.13	100	lf	100	113.4
HP upper drains	SA-106-B	2	80	1.13	100	lf	100	113.4
IP piping and supports					**700**	**lf**		**957**
IP continuous blowdown to continuous blowdown tank	SA-106-B	1.5	80	1.68	120	lf	120	201.5
IP-50 SPHTR 1 to SPHTR 2 drain	SA-106-B	2	80	1.27	80	lf	80	101.8
IP-13-A sparging steam header drain	SA-106-B	2	80	2.76	20	lf	20	55.3
IP economizer drains	SA-106-B	1	80	1.38	140	lf	140	193.9
2"-IP-51	SA-106-B	2	80	1.17	120	lf	120	140.7
IP-16-D drain	SA-106-B	2	80	1.87	20	lf	20	37.5
IP-51 blowdown piping	SA-106-B	2	80	1.13	100	lf	100	113.4
IP/RH upper drains	SA-106-B	2	80	1.13	100	lf	100	113.4
LP piping and supports					**420**	**lf**		**580**
LP-50 SPHTR 1 to SPHTR 2 drain	SA-106-B	2	80	1.02	140	lf	140	142.4
LP-22-A sparging steam drain	SA-106-B	2	80	2.76	20	lf	20	55.3
LP-22-C drain	SA-106-B	2	80	3.06	20	lf	20	61.2
LP-26-A blow off tank drain	SA-106-B	2	80	2.96	20	lf	20	59.2
LP-51 upper blowdown piping	SA-106-B	2	80	1.13	100	lf	100	113.4
LP boot drain	SA-106-B	2	80	1.77	20	lf	20	35.5
LP upper drains	SA-106-B	2	80	1.13	100	lf	100	113.4
RP piping and supports					**385**	**lf**		**604**
RP-50 RH desuperheater spray water	SA-106-B	1.5	80	1.79	65	lf	65	116.2
RP-51 reheater 1−2 drain	SA-106-B	2	80	1.67	40	lf	40	66.8
1-1/2" RP-50	SA-106-B	1.5	80	1.76	60	lf	60	105.4
CRH boot drain	SA-106-B	2	80	1.77	20	lf	20	35.5
HRH boot drain	SA-106-B	2	80	1.77	20	lf	20	35.5
Preheater drains	SA-106-B	2	80	1.36	180	lf	180	244.7
Blowdown piping					**680**	**lf**		**880**
HD-D'-N intermittent blow off	SA-106-B	1.5	80	1.13	120	lf	120	135.5
HD-01-M cascading blowdown to IP drum	SA-106-B	1.5	80	2.06	120	lf	120	247.0
ID-01-M intermittent blow off	SA-106-B	1.5	80	1.13	120	lf	120	135.5
LD-01-N intermittent blow off	SA-106-B	1.5	80	1.13	120	lf	120	135.5
HD-01-N intermittent blow off	SA-106-B	2	80	1.13	100	lf	100	113.4
LD-03-G BD tank to BD tank	SA-106-B	2	80	1.13	100	lf	100	113.4
Instrumentation					**160**	**lf**		**1044.0**
Casing instrumentation	SA-106-B	1.5	80	6.53	**160**	lf	160	**1044.0**

6.11 Heat recovery steam generator—risers and down comers

	Qty		PF	
	n_i	Unit	$n_i MH_a$	MH/LF
	795	lf	3827	4.81
HP Drum down comer	242	lf	2115	8.74
HP evap to HP drum risers	200	lf	832	4.16
IP drum down comer	125	lf	196	1.57
IP evap to IP drum risers	36	lf	216	6.00
LP drum down comer	132	lf	168	1.27
LP evap to LP drum risers	60	lf	300	5.00
6.11.1 HP, IP, and LP drums risers and down comers	**795**	**lf**		**3827**

HP drum							
Down comer				**242**	**lf**		**2115**
			MH	Qty	Unit	Qty	PF
Line no.	Material	Size	Sch/Thk	MH_a	n_i		n_i	$n_i MH_a$

Line no.	Material	Size	Sch/Thk	MH_a	n_i	Unit	n_i	$n_i MH_a$
HP-07 down comer	SA-106-B	26	2.5	9.28	192	lf	192	1782.5
HP-07 down comer	SA-106-C	16	2	7.79	40	lf	40	311.4
HP-07 down comer		4		2.16	10	lf	10	21.6
HP evap to HP drum risers					200	lf		832
AHR11	SA-106-C	8	140	4.16	10	lf	10	41.6
AHR12	SA-106-C	8	140	4.16	10	lf	10	41.6
AHR13	SA-106-C	8	140	4.16	10	lf	10	41.6
BHR11	SA-106-C	8	140	4.16	10	lf	10	41.6
BHR12	SA-106-C	8	140	4.16	10	lf	10	41.6
BHR13	SA-106-C	8	140	4.16	10	lf	10	41.6
CHR11	SA-106-C	8	140	4.16	10	lf	10	41.6
CHR12	SA-106-C	8	140	4.16	10	lf	10	41.6
CHR13	SA-106-C	8	140	4.16	10	lf	10	41.6
AHR21	SA-106-C	8	140	4.16	10	lf	10	41.6
AHR22	SA-106-C	8	140	4.16	10	lf	10	41.6
BHR21	SA-106-C	8	140	4.16	10	lf	10	41.6
BHR22	SA-106-C	8	140	4.16	10	lf	10	41.6
CHR21	SA-106-C	8	140	4.16	10	lf	10	41.6
CHR22	SA-106-C	8	140	4.16	10	lf	10	41.6
AHR31	SA-106-C	8	140	4.16	10	lf	10	41.6
AHR32	SA-106-C	8	140	4.16	10	lf	10	41.6
AHR33	SA-106-C	8	140	4.16	10	lf	10	41.6
BHR31	SA-106-C	8	140	4.16	10	lf	10	41.6
BHR32	SA-106-C	8	140	4.16	10	lf	10	41.6
IP drum					125	lf		196
IP-05 down comer	SA-106-B	10	40	1.52	114	lf	114	173.3
IP-05 down comer	SA-106-B	5	40	1.98	9	lf	9	17.9
IP-05 down comer		4		2.40	2	lf	2	4.8
IP evap to IP drum risers					36	lf		216
AIR11	SA-106-B	10	40	6.00	6	lf	6	36.0
AIR12	SA-106-B	10	40	6.00	6	lf	6	36.0
BIR11	SA-106-B	10	40	6.00	6	lf	6	36.0
BIR12	SA-106-B	10	40	6.00	6	lf	6	36.0

(*Continued*)

(Continued)

				Qty		PF		
				n_i	Unit	$n_i MH_a$	MH/LF
CIR21	SA-106-B	10	40	6.00	6	lf	6	36.0
CIR22	SA-106-B	10	40	6.00	6	lf	6	36.0
LP drum								
Down comer					**132**	**lf**		**168.0**
LP-09 down comer	SA-106-B	10	40	1.24	110	lf	110	136.5
LP-09 down comer	SA-106-B	6	40	1.34	20	lf	20	26.7
LP-09 down comer		4		2.40	2	lf	2	4.8
LP evap to LP drum risers					**60**	**lf**		**300**
ALR11	SA-106-B	12	80	5.00	10	lf	10	50.0
ALR12	SA-106-B	12	80	5.00	10	lf	10	50.0
BLR11	SA-106-B	12	80	5.00	10	lf	10	50.0
BLR12	SA-106-B	12	80	5.00	10	lf	10	50.0
CLR11	SA-106-B	12	80	5.00	10	lf	10	50.0
CLR12	SA-106-B	12	80	5.00	10	lf	10	50.0

6.12 Heat recovery steam generator—field trim piping

	Qty		PF	
	n_i	Unit	$n_i MH_a$	MH/LF
	1570	**lf**	**3341**	**2.1**
HP remote steam drum	80	lf	392	4.9
IP remote steam drum	80	lf	310	3.9
LP remote steam drum	80	lf	310	3.9
Instrumentation	1240	lf	2114	1.7
Atmospheric blow off tank	45	lf	128	2.9
Flash separator	45	lf	86	1.9

6.12.1 Heat recovery steam generator field trim piping—HP, IP, LP remote drums

				MH	Qty	Unit	Qty	PF
Line no.	Material	Size	Sch/Thk	MH_a	n_i		n_i	$n_i MH_a$
HRSG—field trim piping					**1570**	**lf**		**3341**
HP remote steam drum					**80**	**lf**		**392.0**
HD-01-K PI-X305	SA-106-B	1	80	4.9	20	lf	20	98.0
HD-01-F LG-X301 w/LT-X301C (top conn)	SA-106-B	1	80	4.9	20	lf	20	98.0
HD-01-G LG-X301 w/LT-X301C (bottom conn)								
HD-01-H LG-X301 (drain)								
HD-01-A LI-X301 w/LT-X310A (top conn)	SA-106-B	1	80	4.9	20	lf	20	98.0

(*Continued*)

(Continued)

Line no.	Material	Size	Sch/Thk	MH MH_a	Qty n_i	Unit	Qty n_i	PF $n_i MH_a$
HD-01-B LI-X301 w/LT-X301A (bottom conn)								
HD-01-C LI-X301 (drain)								
HD-01-D LT-X301B w/PT-X305B (top conn)	SA-106-B	1	80	4.9	20	lf	20	98.0
HD-01-E LT-X301B w/PT-X305B (bottom conn)								
IP remote steam drum					**80**	**lf**		**310**
ID-01-F LG-X201 w/LT-X201C (top conn)	SA-106-B	1	80	6.44	20	lf	20	128.8
ID-01-G LG-X201 w/LT-X201C (bottom conn)								
ID-01-H LG-X201 (drain)								
ID-01-A LI-X201 w/LT-X210A (top conn)	SA-106-B	1	80	4.9	20	lf	20	98.0
ID-01-B LI-X201 w/LT-X201A (bottom conn)								
ID-01-C LI-X201 (drain)								
ID-01-D LT-X201B w/PT-X203B (top conn)	SA-106-B	1	80	2.24	20	lf	20	44.8
ID-01-E LT-X201B w/PT-X203B (bottom conn)								
ID-01-K PI-X203	SA-106-B	1	80	1.94	20	lf	20	38.8
LP remote steam drum					**80**	**lf**		**310**
LD-01-F LG-X101 w/LT-X101C (top conn)	SA-106-B	1	80	6.44	20	lf	20	128.8
LD-01-G LG-1201 w/LT-X101C (bottom conn)								
LD-01-H LG-X101 (drain)								
LD-01-A LI-X101 w/LT-X110A (top conn)	SA-106-B	1	80	4.9	20	lf	20	98.0
LD-01-B LI-X101 w/LT-X101A (bottom conn)								
LD-01-C LI-X101 (drain)								
LD-01-D LT-X101B w/PT-X103B (top conn)	SA-106-B	1	80	2.24	20	lf	20	44.8
LD-01-E LT-X101B w/PT-X103B (bottom conn)								
LD-01-N PI-X106	SA-106-B	1	80	1.94	20	lf	20	38.8
Instrumentation					**1240**	**lf**		**2114**
IP instrumentation	SA-106-B	0.8	80	1.7	420	lf	420	715.0
HP instrumentation	SA-106-B	0.8	80	1.75	420	lf	420	734.6
HP instrumentation		0.8	80	1.66	400	lf	400	664.2
Atmospheric blow off tank					**45**	**lf**		**128**
LD-04-A LG-X103 (top conn)	SA-106-B	1	80	3.91	20	lf	20	78.1
LD-04-B LG-X103 (bottom conn)								
PI-X114	SA-106-B	1	80	1.94	20	lf	20	38.8
TI-X105	SA-106-B	1	80	2.3	5	lf	5	11.5
Flash separator					**45**	**lf**		**86**
LD-04-A LG-X103 (top conn)	SA-106-B	1	80	2.61	20	lf	20	52.2
LD-04-B LG-X103 (bottom conn)								
PI-X114	SA-106-B	1	80	1.28	20	lf	20	25.6
TI-X105	SA-106-B	1	80	1.6	5	lf	5	8.0

6.13 SP-01 AIG piping

	Qty		PF	
	n_i	Unit	$n_i\,MH_a$	MH/LF
	1332	**lf**	**2262**	**1.70**
SP-01 AIG manifold connecting piping	73	lf	119	1.63
Inner connect pipe	1259	lf	2143	1.70

SP-01 AIG piping						1332	lf		2262
				MH	Qty	Unit	Qty	PF
Line no.	Material	Size	Sch/Thk	MH_a	n_i		n_i	$n_i\,MH_a$
SP-01 AIG manifold connecting piping	SA-106-B	10	40	1.63	73	lf	73	118.9
Inner connect pipe					1259	lf		2143
SP-02 AIG piping	SA-106-B	3	40	1.88	28	lf	28	52.8
SP-03 AIG piping	SA-106-B	3	40	1.10	21	lf	21	23.1
SP-04 AIG piping	SA-106-B	3	40	1.18	15	lf	15	17.7
SP-05 AIG piping	SA-106-B	3	40	1.36	10	lf	10	13.6
SP-06 AIG piping	SA-106-B	3	40	1.84	30	lf	30	55.2
SP-07 AIG piping	SA-106-B	3	40	1.84	37	lf	37	68.1
SP-08 AIG piping	SA-106-B	3	40	1.75	43	lf	43	75.4
SP-09 AIG piping	SA-106-B	3	40	1.69	49	lf	49	82.8
SP-10 AIG piping	SA-106-B	3	40	1.71	56	lf	56	95.5
SP-11 AIG piping	SA-106-B	3	40	1.71	62	lf	62	105.7
SP-12 AIG piping	SA-106-B	3	40	1.71	68	lf	68	116.0
SP-13 AIG piping	SA-106-B	3	40	1.72	75	lf	75	128.7
SP-14 AIG piping	SA-106-B	3	40	1.68	81	lf	81	136.2
SP-15 AIG piping	SA-106-B	3	40	1.79	33	lf	33	59.1
SP-16 AIG piping	SA-106-B	3	40	1.91	27	lf	27	51.5
SP-17 AIG piping	SA-106-B	3	40	1.94	20	lf	20	38.9
SP-18 AIG piping	SA-106-B	3	40	1.99	15	lf	15	29.8
SP-19 AIG piping	SA-106-B	3	40	1.74	36	lf	36	62.7
SP-20 AIG piping	SA-106-B	3	40	1.77	42	lf	42	74.1
SP-21 AIG piping	SA-106-B	3	40	1.70	48	lf	48	81.6
SP-22 AIG piping	SA-106-B	3	40	1.64	55	lf	55	90.2
SP-23 AIG piping	SA-106-B	3	40	1.67	61	lf	61	101.9
SP-24 AIG piping	SA-106-B	3	40	1.67	67	lf	67	111.9
SP-25 AIG piping	SA-106-B	3	40	1.63	74	lf	74	120.6
SP-26 AIG piping	SA-106-B	3	40	1.65	80	lf	80	132.0
SP-27 AIG piping	SA-106-B	3	40	1.65	87	lf	87	143.6
SP-28 AIG piping	SA-106-B	10	40	1.90	39	lf	39	74.2

6.14 Excel double-flow STG installation estimate man-hours

	MH	BM	IW	MW	PF
	23,363	1636	1335	7449	12,943
Excel spreadsheet summary STG vendor piping	7014				7014
Excel spreadsheet summary STG piping	5928				5928
Excel spreadsheet summary STG equipment	10,420	1636	1335	7449	

6.15 Excel double-flow STG equipment installation estimate

n_i MH_a	BM	IW	MW	
Excel double-flow STG installation man-hours	**10,420**	**1636**	**1335**	**7449**
Estimate—centerline equipment	3729	306	195	3228
Estimate HP/IP turbine and front standard	347		116	231
Final turbine assembly and alignment	2106			2106
Generator acoustic enclosure	1447	350	504	593
Generator installation	1811		520	1291
Lube oil system	400	400		
Hydraulic power oil system	580	580		

6.16 Excel double-flow STG installation estimate

Description	MH	Qty	Unit	Qty				
	MH_a	n_i		n_i	n_i MH_a	BM	IW	MW
6.16.1 Estimate—centerline equipment					3729	306	195	3228
Layout foundations	0.12	3120.0	SF	3120.0	374			374
Set lower exhaust hood								
Clean protective coating	0.25	104.0	LF	104.0	26			26
Weld LO supply and drain down comers	6.00	2.0	EA	2.0	12			12
Bolt connections—mid-standard	0.50	12.0	EA	12.0	6			6
Fit mid-standard to sole plates	2.00	4.0	EA	4.0	8			8
Install and grout retaining strips	0.50	104.0	LF	104.0	52			52
Set lower turbine and exhaust hood	1.02	176.9	ton	176.9	180		180	
Clean and set turning gear standard	80.00	1.0	EA	1.0	80			80
Move exhaust hood	80.00	1.0	EA	1.0	80			80
Rough alignment	1.42	176.9	ton	176.9	251			251
Seal weld mid-standard	0.50	30.0	LF	30.0	15		15	
Install lower inner casing								
Rig and install inner casing	2.56	52.4	ton	52.4	134			134
Install lower half packing casing	20.00	2.0	EA	2.0	40			40
Rough alignment	1.42	52.4	ton	52.4	74			74
Install upper inner casing								
Rig and install upper casing	2.56	56.4	ton	56.4	144			144
Bolt connection—inner casing	0.50	112.0	EA	112.0	56			56
Final alignment	2.10	56.4	ton	56.4	118			118

(Continued)

(Continued)

Description	MH	Qty	Unit	Qty				
	MH$_a$	n$_i$		n$_i$	n$_i$ MH$_a$	BM	IW	MW
Crossover flange level check	8.00	1.0	EA	1.0	8			8
Install upper half exhaust hood								
Install upper exhaust hood halves	2.56	77.0	ton	77.0	197			197
Rough alignment	1.42	77.0	ton	77.0	109			109
Bolt connection—vertical joint	0.50	60.0	EA	60.0	30			30
Close exhaust hood								
Bolt connection—exhaust hood	0.50	216.0	EA	216.0	108			108
Final alignment	2.10	176.9	ton	176.9	371			371
Exhaust hood centerline sole plate								
Adjust left and right-side anchor blocks	20.00	8.0	EA	8.0	160			160
Weld anchor blocks exhaust hood	10.00	8.0	EA	8.0	80			80
Tops-on exhaust hood alignment								
Final alignment	2.10	77.0	ton	77.0	162			162
Turning gear standard								
Final alignment	2.10	52.4	ton	52.4	110			110
Weld exhaust hood to condenser								
Weld exhaust hood condenser neck	1.50	104.0	LF	104.0	156	156		
Alignment during welding	2.10	52.4	ton	52.4	110	110		
Install condenser joint shield	40.00	1.0	EA	1.0	40	40		
Final grout sole plates								
Clean area and install grout sole plates	16.00	16.0	EA	16.0	256			256
Final tops-on alignment								
Final alignment	2.10	52.4	ton	52.4	110			110
Fit exhaust hood keys								
Fit internal exhaust hood keys	1.00	40.0	EA	40.0	40			40
Set and fit bearing rings								
Set T2, T3, and T4 bearing rings	10.00	3.0	EA	3.0	30			30
6.16.2 Estimate HP/IP turbine and front standard					347		116	231
Front standard base plate check								
Set fixators	3.50	4.0	EA	4.0	14			14
Clean and set front standard plate	20.00	1.0	EA	1.0	20			20
Front standard base plate check								
Rough alignment	1.42	32.0	ton	32.0	45			45
Final alignment	2.10	32.0	ton	32.0	67			67
Set HP/IP turbine assembly								
Set assembled HP/IP turbine	1.70	59.1	ton	59.1	100		100	
Remove turbine end shipping brace	4.00	4.0	EA	4.0	16		16	
Rough alignment	1.42	59.1	ton	59.1	84			84
Final turbine assembly and alignment					2106			2106
Upper exhaust hood—inner casing								
Remove dowels from exhaust hood	0.55	216.0	EA	216.0	120			120
Rig and remove upper exhaust hood	1.56	77.0	ton	77.0	120			120
Remove upper—inner casing								

(Continued)

(Continued)

6.16.2 Estimate HP/IP turbine and front standard					347	116	231
Loosen and remove dowels and bolts	0.50	112.0	EA	112.0	56		56
Rig and remove upper inner casing	2.56	56.4	ton	56.4	144		144
Final tops-off alignment							
Alignment (T3 and T4 inboard oil deflector)	2.10	13.8	ton	13.8	29		29
Lower half low-pressure diaphragms							
Clam protective coating	0.25	79.0	LF	79.0	20		20
Rig and install diaphragms	14.00	12.0	EA	12.0	168		168
Rough alignment	1.42	5.8	ton	5.8	8		8
Number 3 and 4 bearings							
Bearings, bearing rings and standard fits	0.25	36.0	LF	36.0	9		9
Install bearings	16.00	3.0	EA	3.0	48		48
Install upper half bearing rings	16.00	3.0	EA	3.0	48		48
Rough alignment	1.42	17.0	ton	17.0	24		24
Low-pressure rotor							
Clam protective coating	0.25	48.0	LF	48.0	12		12
Rig and install low-pressure rotor	1.65	84.9	ton	84.9	140		140
Install low-pressure oil deflectors							
Clam protective coating	0.25	36.0	LF	36.0	9		9
Install lower half oil deflectors	10.00	12.0	EA	12.0	120		120
Upper half pressure bearings and rings							
Install upper half bearings and rings	20.00	3.0	EA	3.0	60		60
Rough alignment	0.04	600.0	LB	600.0	24		24
Upper half low-pressure diaphragms							
Clam protective coating	0.25	79.0	LF	79.0	20		20
Rig and install diaphragms	14.00	12.0	EA	12.0	168		168
Rough alignment	1.42	5.8	ton	5.8	8		8
Upper half casing							
Rig and install the upper inner casing	2.56	56.4	ton	56.4	144		144
Install horizontal joint shields	0.50	120.0	LF	120.0	60		60
Final alignment	2.10	56.4	ton	56.4	118		118
Final turbine assembly and alignment							
Upper half exhaust casing							
Rig and install the upper exhaust hood	1.56	77.0	ton	77.0	120		120
Rough alignment	1.42	77.0	ton	77.0	109		109
Bolt connection—vertical joint	0.50	60.0	EA	60.0	30		30
Install atmosphere relief diaphragm	16.00	2.0	EA	2.0	32		32
HP/IP—LP assembly							
Set T2 lower half bearings	20.00	1.0	EA	1.0	20		20
Rig and lift generator end of rotor	1.60	20.1	ton	20.1	32		32
Remove shipping brackets	4.00	4.0	EA	4.0	16		16
Install upper T1 and T2 bearings	20.00	2.0	EA	2.0	40		40
Rough alignment	1.42	20.1	ton	20.1	29		29

6.16.3 Generator acoustic enclosure					1447	350	504	593
Front standard base plate								
Grout front standard base plate	1.40	43.0	SF	43.0	60			60
Horn drop check								
Install/remove generator end rotor support	40.00	1.0	EA	1.0	40			40
Final alignment HP/IP rotor								
Final alignment	2.10	105.0	ton	105.0	220			220
"A" coupling								
Clean coupling and associated hardware	0.25	32.0	LF	32.0	8			8
Assemble, fit, and dowel "A" coupling guard	30.00	1.0	EA	1.0	30			30
Rough alignment	1.42	0.5	ton	0.5	1			1
Install and fit "A" coupling covers	10.00	2.0	EA	2.0	20			20
Install and fit upper half oil deflectors	10.00	2.0	EA	2.0	20			20
Install HP/IP centerline keys								
Machine and install center keys	2.58	40.0	EA	40.0	103			103
Rough alignment	1.42	1.0	ton	1.0	1			1
Install crossover								
Remove shipping skid and coatings	0.25	32.0	LF	32.0	8	8		
Remove expansion joint locking device	20.00	1.0	EA	1.0	20	20		
54" diameter fit up and weld—HP/LP turbine	1.15	54.0	DI	54.0	62	62		
54" diameter fir up and weld—assembly	1.15	104.0	DI	104.0	120	120		
Install 4" diameter crossover pipe	3.64	32.0	LF	32.0	116	116		
54" diameter flange boltup	0.45	54.0	DI	54.0	24	24		
Install and align turning gear								
Install turning gear to rotor	40.00	1.0	EA	1.0	40			40
Final alignment	2.10	13.7	ton	13.7	29			29
Dowel turning gear to standard	1.20	16.5	EA	16.5	20			20
Turbine generator sections								
Left lower quarter section	40.00	1.0	EA	1.0	40		40	
Right lower quarter section	40.00	1.0	EA	1.0	40		40	
Middle quarter section	40.00	2.0	EA	2.0	80		80	
End half section	40.00	2.0	EA	2.0	80		80	
Bottom plate	40.00	1.0	EA	1.0	40		40	
Doors	8.00	4.0	EA	4.0	32		32	
Field joints	16.00	12.0	EA	12.0	192		192	

6.16.4 Generator installation					1811		520	1291
Generator terminal compartment								
Assemble cribbing and set compartment	120.0	1.0	EA	1.0	120		120	
Rigging and moving generator								
Clean and install trunnions	10.0	8.0	EA	8.0	80			80
Rough alignment stator								

(Continued)

(Continued)

						1811	520	1291
6.16.4 Generator installation								
Rig, lift, and move and set generator	1.4	259.5	ton	259.5	360	360	
Remove and install generator outer shields	20.0	2.0	EA	2.0	40		40	
Install generator journal bearings	10.0	4.0	EA	4.0	40			40
Boltup outer end shield and dowel	1.0	48.0	EA	48.0	49			49
Install hydrogen seal rings	25.0	1.0	EA	1.0	25			25
Install and align oil deflectors	10.0	12.0	EA	12.0	120			120
Set and rough alignment stator	1.4	185.5	ton	185.5	263			263
Final generator alignment								
Final alignment	2.1	259.5	ton	259.5	545			545
Grout sole plate	16.0	8.0	EA	8.0	128			128
LP turbine generator								
Install studs and align coupling guard	40.0	1.0	EA	1.0	40			40
Estimate—LO system								
Lube oil system						400	400	
Set lube oil tank	60.0	1.0	EA	1.0	60	60	
Align lube oil pumps	20.0	3.0	EA	3.0	60	60	
Install lube oil piping	2.2	100.0	LF	100.0	220	220	
Install lube oil conditioner	60.0	1.0	EA	1.0	60	60	
Estimate—hydraulic power oil system								
Hydraulic power oil system					0.0	580	580	
Set hydraulic power unit	80.0	1.0	EA	1.0	80	80	
Align hydraulic power pumps	20.0	3.0	EA	3.0	60	60	
Install hydraulic power piping	2.2	200.0	LF	200.0	440	440	

6.17 STG vendor piping

	Qty		PF	
	n_i	Unit	n_i MH$_a$	MH/LF
	3040	**lf**	**7014**	**2.3**
Spray water system	60	lf	75	1.25
Gland steam system (15Mo3)	840	lf	1369	1.63
Leak off steam system (15Mo3)	320	lf	**440**	1.37
Lube oil supply	300	lf	456	1.52
Lube oil return	300	lf	633	2.11
Hydrostatic oil, oil purification	200	lf	322	1.61
Jacking oil	300	lf	916	3.05
Oil flushing	20	lf	20	1.00
External hydraulic oil	700	lf	1003	1.43
Hydro			540	
Clean piping			1240	

6.17.1 STG—vendor piping sheet 1

				MH	Qty	Unit	Qty	PF
					3040	lf		7014
Line no.	Material	Size	Sch/Thk	MH_a	n_i		n_i	$n_i MH_a$
Spray water system	SA-106-B	2	80	1.25	**60**	lf	60	**75**
Gland steam system (15Mo3)					**840**	lf		**1369**
Pipe	15Mo3	6	80	1.98	140	lf	140	278
Pipe	15Mo3	2.5	80	1.20	80	lf	80	96
Pipe	15Mo3	5	80	1.37	40	lf	40	55
Pipe	15Mo3	4	80	1.49	40	lf	40	60
Pipe	15Mo3	6	80	1.67	60	lf	60	100
Pipe	15Mo3	2	80	1.20	60	lf	60	72
Pipe	15Mo3	4	80	1.44	40	lf	40	58
Pipe	15Mo3	6	80	2.15	140	lf	140	302
Pipe	15Mo3	6	80	2.74	40	lf	40	110
Pipe	15Mo3	2	80	1.20	140	lf	140	168
Pipe	15Mo3	2	80	1.20	20	lf	20	24
Pipe	15Mo3	2	80	1.20	40	lf	40	48
Leak off steam system (15Mo3)					**320**	lf		**440**
Pipe	15Mo3	2	80	1.20	40	lf	40	48
Pipe	15Mo3	2	80	1.20	40	lf	40	48
Pipe	15Mo3	3	80	1.25	100	lf	100	125
Pipe	15Mo3	3	80	1.18	40	lf	40	47
Pipe	15Mo3	4	80	1.21	40	lf	40	48
Pipe	15Mo3	6	80	1.80	40	lf	40	72
Pipe	15Mo3	6	80	2.56	20	lf	20	51
Lube oil supply					**300**	lf		**456**
Pipe-304L SS	304L SS	6	S10S	1.83	20	lf	20	37
Pipe-304L SS	304L SS	4	S10S	1.82	20	lf	20	36
Pipe-304L SS	304L SS	2	S40S	1.27	40	lf	40	51
Pipe-304L SS	304L SS	1	S40S	1.19	20	lf	20	24
Pipe-304L SS	304L SS	4	S10S	1.48	80	lf	80	119
Pipe-304L SS	304L SS	2	S40S	1.58	20	lf	20	32
Pipe-304L SS	304L SS	2	S40S	1.58	20	lf	20	32
Pipe-304L SS	304L SS	4	S10S	1.73	40	lf	40	69
Pipe-304L SS	304L SS	2	S40S	1.45	40	lf	40	58

6.17.2 STG—vendor piping sheet 2

Lube oil return					300	lf		633
Pipe-304L SS	304L SS	12	S10S	2.80	60	lf	60	168
Pipe-304L SS	304L SS	6	S10S	1.47	20	lf	20	29
Pipe-304L SS	304L SS	5	S10S	1.02	20	lf	20	20
Pipe-304L SS	304L SS	6	S10S	1.72	40	lf	40	69
Pipe-304L SS	304L SS	5	S10S	2.34	20	lf	20	47
Pipe-304L SS	304L SS	5	S10S	2.74	20	lf	20	55
Pipe-304L SS	304L SS	10	S10S	2.84	40	lf	40	114
Pipe-304L SS	304L SS	5	S10S	1.60	20	lf	20	32

(Continued)

(Continued)

Lube oil return						300	lf		633
Pipe-304L SS	304L SS	4	S10S	1.50	20	lf	20	30
Pipe-304L SS	304L SS	4	S10S	2.07	20	lf	20	41
Pipe-304L SS	304L SS	2.5	S10S	1.42	20	lf	20	28
Hydrostatic oil, oil purification					**200**	**lf**		**322**
Pipe-304L SS	304L SS	1.25	S40S	1.50	40	lf	40	60
Pipe-304L SS	304L SS	2	S40S	1.65	40	lf	40	66
Pipe-304L SS	304L SS	2	S40S	1.62	40	lf	40	65
Pipe-304L SS	304L SS	2.5	S10S	1.64	80	lf	80	131
Jacking oil					**300**	**lf**		**916**
Pipe-304L SS	304L SS	20	S10S	8.19	60	lf	60	491
Pipe-304L SS	304L SS	12	S10S	1.99	120	lf	120	238
Pipe-304L SS	304L SS	2.5	S10S	1.55	120	lf	120	186
Oil flushing					**20**	**lf**		**20**
Rubberized hose	304L SS	0.75	S40S	1.00	20	lf	20	20
External hydraulic oil					**700**	**lf**		**1003**
Pipe-SS	304L SS	3	S10S	2.01	220	lf	220	441
Pipe-SS	304L SS	2	S40S	1.60	40	lf	40	64
Pipe-SS	304L SS	2	S40S	1.15	240	lf	240	276
Pipe-SS	304L SS	1.5	S40S	1.11	200	lf	200	222
Hydro								**540**
Clean piping								**1240**

6.18 STG piping

	Qty			PF		
	n_i	Unit	MH_a	$n_i MH_a$	MH/LF
	3480	**lf**		**5928**	**1.7**
Condensate	340	lf	1.4	476	1.4
Auxiliary steam	1100	lf	1.6	1760	1.6
Steam turbine drain	140	lf	1.5	210	1.5
Steam drain	1080	lf	1.5	1620	1.5
Low-pressure steam	320	lf	1.6	512	1.6
High-pressure steam	280	lf	2.2	616	2.2
Hot reheat steam	80	lf	2.4	192	2.4
Trim	20	lf	1.5	30	1.5
Boiler feedwater	120	lf	2	240	2

6.19 Excel F class CTG installation estimate man-hours

	MH	BM	IW	MW	PF
Excel spreadsheet summary CTG equipment	13,260	7231	3837	2192	
Excel spreadsheet summary CTG vendor piping	14,634				14,634
Excel spreadsheet summary CTG equipment and piping	27,894	7231	3837	2192	14,634

6.20 Summary F class CTG installation man-hour estimate

	$n_i MH_a$	BM	IW	MW
F class CTG equipment installation estimate	13,260	7231	3837	2192
Estimate—centerline equipment	2463		271	2192
Modules and skids	815		815	
Place equipment	202		202	
Generator terminal enclosure	180		180	
Turbine enclosure, grating, and walkways	1891		1891	
Turbine load compartment	199		199	
Install enclosure to base	280		280	
Inlet filter house	1730	1730		
Inlet filter house transition duct	337	337		
Support steel	1247	1247		
Inlet ductwork	1311	1311		
Inlet plenum	841	841		
Exhaust duct	374	374		
Exhaust duct/stack	1391	1391		

6.21 Excel F class CTG installation estimate

Description	MH	Qty	Unit	Qty				
	MH_a	n_i		n_i	$n_i MH_a$	BM	IW	MW
6.21.1 Estimate—centerline equipment					2463		271	2192
Set fixators	3.50	16.0	EA	16.0	56			56
Grout fixators	9.20	16.0	EA	16.0	147			147
Rig, lift, and set turbine down onto fixators	0.50	188.5	ton	188.5	94		94	
Install load coupling to turbine	9.25	2.5	ton	2.5	23			23
Rig, lift, and set generator	0.50	280.0	ton	280.0	140		140	
Install collector assembly	9.25	10.0	ton	10.0	93			93
Weld radiation shields and install thermocouple	1.35	27.0	EA	27.0	36		36	
Rig and set exhaust diffuser	9.25	14.1	ton	14.1	131			131
Rough alignment, centerline equipment	1.42	495.1	ton	495.1	703			703
Generator-turbine final alignment	2.10	495.1	ton	495.1	1040			1040
Modules and skids					815		815	
Set accessory package	3.20	37.5	ton	37.5	120		120	

(*Continued*)

(Continued)

Description	MH	Qty	Unit	Qty				
	MH$_a$	n$_i$		n$_i$	n$_i$ MH$_a$	BM	IW	MW
Set liquid fuel and atomizing air skid	4.80	30.0	ton	30.0	144		144	
Set electric control compartment	4.80	23.0	ton	23.0	110		110	
Water injection skid	4.80	15.0	ton	15.0	72		72	
Fire protection skid	5.20	10.0	ton	10.0	52		52	
Set water cooling module	3.20	49.8	ton	49.8	159		159	
Set bus accessory compartment	5.20	7.5	ton	7.5	39		39	
Set liquid fuel forwarding skid	5.20	5.9	ton	5.9	31		31	
Air processing skid	5.20	1.3	ton	1.3	7		7	
Set water wash skid	4.80	10.9	EA	10.9	52		52	
Set cooling fan module	5.20	5.5	lot	5.5	29		29	
Place equipment					202	0	202	0
Set hydrogen dryer and purge assembly	30.00	1.0	lot	1.0	30		30	
Set liquid detectors	20.00	1.0	EA	1.0	20		20	
Install batteries (pallet 1, 2, 3)	20.00	1.0	EA	1.0	20		20	
Install air conditioners	42.00	1.0	EA	1.0	42		42	
Set drain collection tank	30.00	1.0	EA	1.0	30		30	
Set load excitation compartment	20.00	1.0	EA	1.0	20		20	
Set DC link reactor	20.00	1.0	EA	1.0	20		20	
Set isolation/excitation transformer	20.00	1.0	EA	1.0	20		20	
Enclosures and barriers								
Generator terminal enclosure					180	0	180	0
Set generator terminal enclosure	14.40	12.5	ton	12.5	180		180	
Estimate—enclosure, grating and walkways								

6.21.2 Turbine enclosure, grating, and walkways					1891	0	1891	0
Roof section	12.40	5.5	ton	5.5	68		68	
Side upper; frame end upper	12.40	6.5	ton	6.5	80		80	
Door	10.00	4.0	EA	4.0	40		40	
Frame side/end upper/lower	12.40	16.9	ton	16.9	210		210	
Panels	24.00	9.2	ton	9.2	221		221	
Make up field joints; roof, frames, and panels	0.75	723.6	LF	723.6	543		543	
Floor unit	24.00	5.3	ton	5.3	127		127	
Support steel	64.00	0.4	ton	0.4	24		24	
Make up field joints; floor units	0.75	773.0	LF	773.0	580		580	
Estimate—turbine load compartment								
Turbine load compartment					199	0	199	0
Left lower quarter section	60.00	0.4	ton	0.4	26		26	
Right lower quarter section	60.00	0.4	ton	0.4	26		26	
Top half section	60.00	0.7	ton	0.7	39		39	
Bottom plate	60.00	0.5	ton	0.5	32		32	
Doors	10.00	2.0	EA	2.0	20		20	
Field joints	0.75	72.6	LF	72.6	54		54	
Install enclosure to base					280	0	280	0
Set enclosure on base	280.00	1.0	EA	1.0	280		280	

6.21.3 Inlet filter house, transition duct, ductwork					1730	1730
Sloped roof	8.40	2.1	ton	2.1	18	18
hopper	20.00	1.0	ton	1.0	20	20
Pipe/tubing	1.94	49.7	LF	49.7	96	96
Hand rails	0.25	21.1	LF	21.1	5	5
Filter cartridges	0.75	1020.0	EA	1020.0	765	765
Filter modules (center)	8.40	15.5	ton	15.5	130	130
Filter modules (left and right hand)	8.40	9.3	ton	9.3	78	78
Walkway assembly (center)	8.40	4.7	ton	4.7	39	39
Walkway assembly (LH and RH)	8.40	3.5	ton	3.5	29	29
D15 weather hoods	8.40	9.9	ton	9.9	83	83
Field joints	0.75	620.6	LF	620.6	465	465
Inlet filter house transition duct					**337**	**337**
Top section	28.00	1.3	ton	1.3	38	38
Floor section	28.00	1.5	ton	1.5	43	43
Side section	28.00	1.0	ton	1.0	27	27
Side section	28.00	0.6	ton	0.6	17	17
Bell section	28.00	1.3	ton	1.3	37	37
Make-up duct field joints	0.75	235.4	LF	235.4	177	177
Support steel					**1247**	**1247**
Erect support steel	14.50	86.0	ton	86.0	1247	1247
Inlet ductwork					**1311**	**1311**
Transition duct	8.40	5.9	ton	5.9	50	50
Expansion joint (kit)	40.00	1.0	kit	1.0	40	40
Elbow duct upper w/trash screen	8.40	8.7	ton	8.7	73	73
Elbow duct lower	8.40	10.9	ton	10.9	91	91
Heating duct	8.40	4.8	ton	4.8	40	40
Silencer duct w/panels	8.40	22.3	ton	22.3	187	187
Empty duct	8.40	3.1	ton	3.1	26	26
Make-up duct field joints	0.75	1072.2	LF	1072.2	804	804

6.21.4 Inlet plenum, exhaust system, duct/stack					841	841
Plenum support structure	40.00	2.2	ton	2.2	86	86
Plenum extension	28.00	0.8	ton	0.8	22	22
Plenum cone (inlet case extension)	28.00	1.6	ton	1.6	44	44
Lower/upper quarter section	28.00	14.9	ton	14.9	417	417
Lot 1, 2, and 3	40.00	2.5	ton	2.5	100	100
Make-up duct field joints	0.75	227.5	LF	227.5	171	171
Exhaust system						
Exhaust duct					**374**	**374**
Upstream upper/lower quadrant	8.40	11.4	ton	11.4	96	96
Downstream upper/lower quadrant	8.40	10.3	ton	10.3	87	87
Downstream expansion joint (kit)	40.00	1.0	EA	1.0	40	40
Support steel (kit)	40.00	1.0	kit	1.0	40	40
Make-up duct field joints	0.75	148.8	LF	148.8	112	112
Estimate—exhaust duct/stack						
Exhaust duct/stack					**1391**	**1391**
Upstream duct transition	8.40	12.0	ton	12.0	110	110
Duct silencer	8.40	28.0	ton	28.0	235	235

(*Continued*)

(Continued)

6.21.4 Inlet plenum, exhaust system, duct/stack					841	841
Downstream duct transition	8.40	6.0	ton	6.0	50	50
Lower stack transition	8.40	4.5	ton	4.5	38	38
Silencer	5.20	24.0	ton	24.0	125	125
Upper stack transition	8.40	4.5	ton	4.5	38	38
Stack segments	8.40	16.0	ton	16.0	134	134
Make-up duct field joints	0.75	687.4	LF	687.4	516	516
Support steel	24.00	6.1	ton	6.1	145	145

6.22 CTG vendor piping

	Qty		PF	
	n_i	Unit	$n_i\,MH_a$	MH/LF
	8860	**lf**	**14,634**	**1.65**
SB supply	4320	lf	5876	1.36
LB supply	1460	lf	2968	2.03
Interconnect	860	lf	1335	1.55
Duct	360	lf	1022	2.84
Unit SB	840	lf	1153	1.37
Unit LB	1020	lf	2279	2.23

6.22.1 CTG—vendor piping sheet 1

						4320	lf		5876
				MH	Qty	Unit		Qty	PF
Line no.	Material	Size	Sch/Thk	MH_a	n_i			n_i	$n_i\,MH_a$
SB supply					**4320**	**lf**			**5876**
FGA fuel gas supply	CS	2″	80	1.20	40	lf		40	48
TGH combustion turbine supply	CS	2″	80	1.20	80	lf		80	96
TGH combustion turbine supply	304 SS	2‴″	S40S	1.50	320	lf		320	480
TGH combustion turbine supply	Galv	2″	40	1.15	1660	lf		1660	1908
FWF demin water storage and supply	316 SS	2″	S40S	3.97	80	lf		80	317.2
WSD potable water supply	CS	2″	80	1.63	220	lf		220	358.8
CGA hydrogen storage supply	CS	2″	80	2.93	60	lf		60	175.5
CAA station air supply	Galv	2″	40	1.34	1860	lf		1860	2492
LB supply					**1460**	**lf**			**2968**
Combustion turbine	CS	2-1/2″	40	1.61	80	lf		80	128.7
Combustion turbine	SS	2-1/2″	S40S	1.40	60	lf		60	84
Combustion turbine	Galv	2-1/2″	40	1.20	220	lf		220	264
Combustion turbine	CS	3″	40	2.93	40	lf		40	117
Wastewater collection and disposal	CS	3″	40	2.08	40	lf		40	83.2
Combustion turbine	CS	4″	40	0.93	340	lf		340	315.9
Fuel gas	CS	6″	40	1.56	20	lf		20	31.2

(Continued)

(Continued)

| | | | | 4320 | lf | | 5876 |
| | | | | MH | Qty | Unit | Qty | PF |
Line no.	Material	Size	Sch/Thk	MH_a	n_i		n_i	$n_i\,MH_a$
Fuel gas	CS	8″	40	2.67	20	lf	20	53.3
Combustion turbine	CS	8″	40	2.77	180	lf	180	497.9
Combustion turbine	SS	8″	S40S	2.15	40	lf	40	85.8
Fuel gas	CS	10″	40	3.92	80	lf	80	313.3
Combustion turbine	CS	10″	40	1.86	100	lf	100	185.9
Combustion turbine	SS	10″	S40S	3.64	40	lf	40	145.6
Fuel gas	CS	12″	40	8.84	20	lf	20	176.8
Combustion turbine	SS	12″	S40S	2.50	180	lf	180	449.8
Pipe cleaning/flushing					1		1	36
Interconnect					**860**	**lf**		**1335**
Water and carbon dioxide	CS	1″	80	2.28	40	lf	40	91
Water and carbon dioxide	CS	2″	80	3.85	100	lf	100	384.8
Drains	CS	2″	80	1.20	240	lf	240	288
Water wash	CS	2″	80	1.20	120	lf	120	144
Fire protection	CS	2″	80	1.13	360	lf	360	405.6
Pipe cleaning/flushing					1		1	22

6.22.2 CTG—vendor piping sheet 2

Duct						360	lf	1022
Inlet air heating duct air	CS	8″			1.70	160	lf	160	272
Venting square duct	Duct	48″			3.75	200	lf	200	750
Pipe cleaning/flushing								
Unit SB						**840**	**lf**	**1153**
MSD no. 2 bearing lube oil feed drain	CS	2″	80	1.17	20	lf	20	23.4
MSD no. 1 bearing lube oil feed drain	CS	2″	80	1.17	20	lf	20	23.4
MSD load compartment LO feed drain/sealing air/lift oil	CS	2″	80	1.35	120	lf	120	162
MSD end of generator LO feed drain/sealing air/lift oil	CS	2″	80	1.50	80	lf	80	120
Control oil	CS	2″	80	1.40	120	lf	120	168
Accessory module MSD LO feed drain/fuel gas	CS	2″	80	1.40	80	lf	80	112
Compressor water wash	CS	2″	80	1.40	80	lf	80	112
Inlet and exhaust drains	CS	2″	80	1.40	80	lf	80	112
Seal oil	CS	2″	80	1.40	80	lf	80	112
Liquid level detector	CS	2″	80	1.40	80	lf	80	112
H_2 control	CS	2″	80	1.20	80	lf	80	96
Unit LB						**1020**	**lf**	**2279**
MSD no. 1 bearing lube oil feed drain	CS	2-1/2″	40	1.30	20	lf	20	26
MSD no. 2 bearing lube oil feed drain	CS	4″	40	2.08	20	lf	20	41.6
Generator hydrogen gas supply	CS	4″	40	1.30	80	lf	80	104
Accessory module MSD LO feed drain/fuel gas	CS	6″	40	1.38	100	lf	100	137.8
MSD no. 2 bearing lube oil feed drain	CS	8″	40	2.18	40	lf	40	87.1
MSD no. 1 bearing lube oil feed drain	CS	8″	40	2.18	40	lf	40	87.1
Air extraction	CS	8″	40	8.50	100	lf	100	850.2
Cooling water to turbine flame scanners	CS	8″	40	1.32	60	lf	60	79.2
Generator cooling water	CS	8″	40	1.20	100	lf	100	120.3
Cooling and sealing air	CS	12″	std	1.35	120	lf	120	161.9

(Continued)

(Continued)

Duct						360	lf		1022
Cooling fan module (exhaust frame blowers)	CS	12″	std	1.34	100	lf	100	133.9
Generator lube oil	CS	12″	std	1.42	100	lf	100	141.7
Cooling fan module	CS	24″	std	4.71	40	lf	40	188.5
Performance monitor tubing	CS			1.00	60	lf	60	60
H_2 supply manifold, dryer, and CO_2 manifold	CS			1.5	40	lf	40	60

6.23 BOP equipment

6.23.1 Balance of plant installation man-hour estimate	6644	1475	1552	3618
	$n_i\,MH_a$	BM	IW	MW
Excel estimate—pumps	3274	141	514	2619
Excel estimate—tanks and vessels	650	650		
Excel estimate—skid-mounted units	468		468	
Excel estimate—heaters	280	280		
Excel estimate—compressors	1568		570	998
Excel estimate—heat exchangers	180	180		
Excel estimate—eyewash/shower	224	224		

6.23.2   Balance of plant equipment sheet 1

Description		MH	Qty	Unit	Qty				
	HP	MH_a	n_i		n_i	$n_i\,MH_a$	BM	IW	MW
Excel estimate—pumps						3274	141	514	2619
Boiler feedwater pump A/B	320.0	1.00	2	ea.	2	640	128		512
HRSG blowdown sump pump A/B	15.0	2.10	2	ea.	2	63	12.6		50.4
Condensate pump A/B/C	75.0	1.60	3	ea.	3	360		72	288
Vacuum pump skid A/B	30.0	2.10	2	ea.	2	126		25.2	100.8
CT area sump pump A	15.0	2.10	1	ea.	1	31.5		6.3	25.2
CT area sump pump A	15.0	2.10	1	ea.	1	31.5		6.3	25.2
HP STG area sump pump A/B	15.0	2.10	2	ea.	2	63		12.6	50.4
HP WT area sump pump A	15.0	2.10	1	ea.	1	31.5		6.3	25.2
HP circulating water pump 1 and 2	50.0	1.80	2	ea.	2	180		36	144
HP potable water well pump	25.0	2.10	1	ea.	1	52.5		10.5	42
Auxiliary cooling water pump	75.0	1.60	1	ea.	1	120		24	96
HP closed cooling water pump A/B/C	250.0	1.30	3	ea.	3	975		195	780
HP STG wastewater injection pump A/B	75.0	1.60	2	ea.	2	240		48	192
Wastewater injection pump A/B	75.0	1.60	2	ea.	2	240		48	192
HP condensate extraction pumps	75.0	1.60	1	ea.	1	120		24	96
Excel estimate—tanks and vessels						650	650		
HRSG blowdown tank (0−5 ton)		60	1	ea.	1	60	60		
Closed cooling water chemical feed tank (0−5 ton)		60	1	ea.	1	60	60		
		60	1	ea.	1	60	60		

(*Continued*)

(Continued)

Description	HP	MH$_a$	n_i	Unit	n_i	n_i MH$_a$	BM	IW	MW
Closed cooling water expansion tank (0–5 ton)									
Separator drains tank (0–5 ton)		60	1	ea.	1	60	60		
Water wash drains tank (UG)		60	1	ea.	1	60	60		
External steam drains tank (0–5 ton)		60	1	ea.	1	60	60		
Fuel gas scrubber drains tank (6–10 ton)		90	1	ea.	1	90	90		
Demineralized water storage tank (11–20 ton)		140	1	ea.	1	140	140		
Potable water bladder tank (0–5 ton)		60	1	ea.	1	60	60		
Excel estimate—skid-mounted units						**468**	**468**		
Injection water filtration skid (0–3000 lb)		28	1	ea.	1	28		28	
Condensate polishing skid (3001–6000 lb)		40	1	ea.	1	40		40	
Fuel gas scrubber skid (6001–15,000 lb)		60	1	ea.	1	60		60	
Wastewater injection pump skid (0–3000 lb)		28	1	ea.	1	28		28	
Circulating water sulfuric acid dosing pump skid (0–3000 lb)		28	1	ea.	1	28		28	
Circulating water corrosion inhibitor dosing skid (0–3000 lb)		28	1	ea.	1	28		28	
Circulating water dispersant dosing pump skid (0–3000 lb)		28	1	ea.	1	28		28	
Service water pump skid (0–3000 lb)		28	1	ea.	1	28		28	
Closed cooling water pump skid (0–3000 lb)		28	1	ea.	1	28		28	
Circulating water sodium hypochlorite dosing pump skid (0–3000 lb)		28	1	ea.	1	28		28	
Circulating water biocide dosing pump skid (0–3000 lb)		28	1	ea.	1	28		28	
Ammonia dosing pump skid (0–3000 lb)		28	1	ea.	1	28		28	
Oxygen dosing pump skid (0–3000 lb)		28	1	ea.	1	28		28	
Rotary screw air compressor and receiver/dryer skid (6001–15,000 lb)		60	1	ea.	1	60		60	

6.23.3 Balance of plant equipment sheet 2

Excel estimate—heaters						280	280		
Condensate preheater (0–500 lb)		30	1	ea.	1	30	30		
CT fuel gas heater (501–1500 lb)		50	1	ea.	1	50	50		
Fuel gas performance heater (1001–2000 lb)		80	1	ea.	1	80	80		
CLG. TWR. Chem tank area ESW. Heater (0–500 lb)		30	1	ea.	1	30	30		
Water treatment bldg. ESW. Heater (0–500 lb)		30	1	ea.	1	30	30		
CLG. TWR. Chem tote area ESW. Heater (0–500 lb)		30	1	ea.	1	30	30		
Sample panel area ESW. Heater (0–500 lb)		30	1	ea.	1	30	30		
Excel estimate—compressors						**1568**		**570**	**998**
Fuel gas compressor A/B 500 HP; 22.6 ton (2 ea.)		624	2	ea.	2	1248		250	998
Gas compressor cooler 1 and 2 (2 ea.)		60	2	ea.	2	120		120	
Air compressor A/B 0–5000 LBS (2 ea.)		100	2	ea.	2	200		200	
Excel estimate—heat exchangers						**180**	**180**		

(*Continued*)

(Continued)

Excel estimate—heaters					280	280
Closed cooling water heat exchanger A/B/C	60	3	ea.	3	180	180
Excel estimate—eyewash/shower					**224**	**224**
CLG. TWR. Chem tank area 1 eyewash/shower	32	1	ea.	1	32	32
CLG. TWR. Chem tank area 2 eyewash/shower	32	1	ea.	1	32	32
CLG. TWR. Chem tank area 3 eyewash/shower	32	1	ea.	1	32	32
Water treatment bldg. area 1 eyewash/shower	32	1	ea.	1	32	32
Water treatment bldg. area 2 eyewash/shower	32	1	ea.	1	32	32
CLG. TWR. Chem tote area eyewash/shower	32	1	ea.	1	32	32
Sample panel area eyewash/shower	32	1	ea.	1	32	32

6.24 Structural steel

6.24.1 Structural steel installation man-hour estimate

	11,837	11,837
	$n_i MH_a$	IW
HRSG utility steel bridge	3804	3804
STG utility bridge steel	3453	3453
STG utility bridge steel inside enclosure	691	691
GSU transformer access platforms	311	311
Stair tower and ladders	464	464
Iso phase support steel	759	759
Dead-end structure	759	759
230 kV Switchyard structure including bus supports	380	380
Transmission line poles	138	138
Grating, handrail, and toe plate	1080	1080

6.24.2 Structural steel estimate

Description	MH	Qty	Unit	Qty		
	MH_a	n_i		n_i	$n_i MH_a$	IW
HRSG utility steel bridge				**204.6**	**3803.8**	**3804**
Light—0–19 lb/ft.	24	11	ton	11	264	264
Medium—20–39 lb/ft.	23	33	ton	33	759	759
Heavy—40–79 lb/ft.	18	105.6	ton	105.6	1900.8	1901
X heavy—80–120 lb/ft.	16	55	ton	55	880	880
STG utility bridge steel				**187**	**3453**	**3453**
Light—0–19 lb/ft.	24	8.4	ton	8.4	201.6	201.6
Medium—20–39 lb/ft.	23	27.9	ton	27.9	641.7	641.7
Heavy—40–79 lb/ft.	18	99	ton	99	1782	1782
X heavy—80–120 lb/ft.	16	51.7	ton	51.7	827.2	827.2
				37	**691**	**691**

(Continued)

(Continued)

Description	MH	Qty	Unit	Qty		
	MH$_a$	n$_i$		n$_i$	n$_i$ MH$_a$	IW
STG utility bridge steel inside enclosure						
Light—0–19 lb/ft.	24	1.7	ton	1.7	40.8	40.8
Medium—20–39 lb/ft.	23	5.6	ton	5.6	128.8	128.8
Heavy—40–79 lb/ft.	18	19.8	ton	19.8	356.4	356.4
X heavy—80–120 lb/ft.	16	10.3	ton	10.3	164.8	164.8
GSU transformer access platforms				**14**	**311**	**311**
Light and medium	23	13.5	ton	13.5	310.5	310.5
Stair Tower and ladders				**14.2**	**464**	**464**
Light and medium	23	10.4	ton	10.4	239.2	239.2
Stair treads	0.55	300	ea.	300	165	165
Ladders	0.30	200	lf	200	60	60
Iso-phase support steel				**33**	**759**	**759**
Light and medium	23	33	ton	33	759	759
Dead-end structure				**33**	**759**	**759**
Light and medium	23	33	ton	33	759	759
230 kV Switchyard structure including bus supports				**16.5**	**380**	**380**
Light and medium	23	16.5	ton	16.5	379.5	379.5
Transmission line poles				**6**	**138**	**138**
Light and medium	23	6	ton	6	138	138
Grating, handrail and toe plate					**1080**	**1080**
2″ Serrated floor grating	0.05	11,200	lf	11,200	560	560
1.5″ Serrated floor grating	0.05	4100	lf	4100	205	205
Handrail and toe plate	0.15	2100	lf	2100	315	315

6.25 Summary underground piping man-hours

	Qty		PF/Laborer	
	n$_i$	Unit	n$_i$ MH$_a$	MH/LF
Underground piping	**8699**	**lf**	**13,913**	**1.60**
Estimate EKG-fuel gas	1130	lf	1766	1.56
Estimate HAN-boiler blowdown	730	lf	580	0.79
Estimate GAC-raw water	850	lf	1236	1.45
Estimate GHC-condensate makeup	1000	lf	547	0.55
Estimate GKB-potable water	900	lf	455	0.51
Estimate GMA-wastewater collection	1068	lf	606	0.57
Estimate EKG-fuel gas	1130	lf	1665	1.47
Estimate UG cicr water piping	1891	lf	7058	3.73

6.25.1 Underground piping					8699	lf		13,913
	Specification			MH	Qty	Unit	Qty	PF/Laborer
Line no.	Material	Size	Sch/Thk	MH_a	n_i		n_i	$n_i MH_a$
Estimate EKG-fuel gas					**1130**	**lf**		**1766**
EKG-fuel gas	30A3S	12	std wt	1.72	610	lf	610	1047
EKG-fuel gas	30A3S	6	std wt	1.40	460	lf	460	644
EKG-fuel gas	30A3S	2	xs	1.25	60	lf	60	75
Estimate HAN-boiler blowdown					**730**	**lf**		**580**
HAN-boiler blowdown	15F51S	8	150# DI-Cement Lined	1.04	180	lf	180	187
HAN-boiler blowdown	15F51S	3	150# DI-Cement Lined	0.66	140	lf	140	93
HAN-boiler blowdown	15F51S	4	150# DI-Cement Lined	2.76	40	lf	40	110
HAN-boiler blowdown	15A1S	4	std wt	0.51	370	lf	370	190
Estimate GAC-raw water					**850**	**lf**		**1236**
GAC-raw water	12J8S	12	150# HDPE	1.45	850	lf	850	1236
Estimate GHC-condensate makeup					**1000**	**lf**		**547**
GHC-condensate makeup	12J8S	4	150# HDPE	0.60	510	lf	510	306
GHC-condensate makeup	12J8S	3	150# HDPE	0.55	200	lf	200	110
GHC-condensate makeup	12J8S	2	150# HDPE	0.45	290	lf	290	131
Estimate GKB-potable water					**900**	**lf**		**455**
GKB-potable water	12J62S	3	150# HDPE	0.55	500	lf	500	275
GKB-potable water	12J62S	2	150# HDPE	0.45	400	lf	400	180
Estimate GMA-wastewater collection					**1068**	**lf**		**606**
GMA-wastewater collection	12J8S	4	150# HDPE	0.51	468	lf	468	239
GMA-wastewater collection	12J8S	4	150# HDPE	0.58	168	lf	168	97
GMA-wastewater collection	12J8S	4	150# HDPE	0.71	49	lf	49	35
GMA-wastewater collection	12J8S	4	150# HDPE	0.60	276	lf	276	166
GMA-wastewater collection	12J8S	4	150# HDPE	0.65	106	lf	106	69
Estimate EKG-fuel gas					**1130**	**lf**		**1665**
EKG-fuel gas	30A3S	12	std wt	1.72	610	lf	610	1047
EKG-fuel gas	30A3S	6	std wt	1.20	460	lf	460	552
EKG-fuel gas	30A3S	2	xs	1.10	60	lf	60	66
Estimate UG cicr water piping					**1891**	**lf**		**7058**
UG cicr water piping cylinder pipe bell end		54-16		3.73	1891	lf	1891	7058

6.26 Summary aboveground piping man-hours

	Qty		PF	
	n_i	Unit	$n_i MH_a$	MH/LF
Aboveground piping	**38,914**	**lf**	**77,741**	**2.0**
Estimate EKG-fuel gas	1160	lf	2362	2.0
Estimating EKT-fuel gas heating	662	lf	1056	1.6
Estimate GAC-raw water	25	lf	126	5.0
Estimate GHC-condensate makeup	645	lf	761	1.2
Estimate GKB-potable water	875	lf	1016	1.2
Estimate GMA-wastewater collection	1370	lf	1479	1.1
Estimate HAN-boiler blowdown	4033	lf	5484	1.4
Estimate HSJ-anhydrous ammonia	354	lf	238	0.7
Estimate LAB-boiler feedwater	1344	lf	2273	1.7
Estimate LBA-high-pressure steam	530	lf	2484	4.7
Estimate LBB-hot reheat steam	473	lf	2269	4.8
Estimate LBC-cold reheat steam	505	lf	1950	3.9
Estimate LBG-auxiliary steam	490	lf	669	1.4
Estimate-LBM-low-pressure steam	462	lf	1229	2.7
Estimate-LCA-condensate	1626	lf	3154	1.9
Estimate LDK-condensate polishing	370	lf	732	2.0
Estimate MA-steam turbine	500	lf	593	1.2
Estimate MAJ-air extraction	320	lf	537	1.7
Estimate MAL-steam turbine drains	2900	lf	2973	1.0
Estimate MB-combustion turbine	508	lf	719	1.4
Estimate PAB-circulating water	710	lf	7759	10.9
Estimate PBN-circulating water chemical feed	1225	lf	1242	1.0
Estimate PCB-service water	1620	lf	2271	1.4
Estimate PGB-closed cooling water	3135	lf	6548	2.1
Estimate QC-cycle chemical feed	880	lf	880	1.0
Estimate QFB-instrument/service air	5600	lf	6447	1.2
Estimate QH-aux boiler	650	lf	1057	1.6
Estimate QJ-nitrogen	500	lf	550	1.1
Estimate QU-sampling and analysis	5442	lf	18,884	3.5

6.27 Aboveground balance of plant piping

6.27.1 Aboveground piping sheet 1

						38,914	lf		77,741
	Specification				MH	Qty	Unit	Qty	PF
Line no.	Material	Size	Sch/Thk	MH_a		n_i		n_i	$n_i MH_a$
Estimate EKG-fuel gas						**1160**	**lf**		**2362**
EKG-fuel gas	30A3S	12	std wt	2.53		170	lf	170	429
EKG-fuel gas	30A3S	8	std wt	2.53		35	lf	35	89

(Continued)

(Continued)

					MH	Qty	Unit	Qty	PF
						38,914	If		77,741
	Specification				MH	Qty	Unit	Qty	PF
Line no.	Material	Size	Sch/Thk	MH_a		n_i		n_i	$n_i MH_a$
EKG-fuel gas	30A3S	6	std wt	7.19		15	If	15	108
EKG-fuel gas	30A3S	2	xs	1.70		250	If	250	425
EKG-fuel gas	30A3S	1.5	xs	1.46		40	If	40	59
EKG-fuel gas	30A3S	1	xs	2.49		20	If	20	50
EKG-fuel gas	30A3S	8	std wt	2.76		100	If	100	276
EKG-fuel gas	30A3S	8	std wt	1.84		100	If	100	184
EKG-fuel gas	30A3S	2	xs	1.76		300	If	300	528
EKG-fuel gas	30A3S	2	xs	1.73		100	If	100	173
EKG-fuel gas	30A3S	1.5	xs	1.13		20	If	20	23
EKG-fuel gas	30A3S	1	xs	1.93		10	If	10	19
Estimating EKT-fuel gas heating						**662**	**If**		**1056**
EKT-fuel gas heating	90A1S	24	1.531	30.1		10	If	10	301
EKT-fuel gas heating	90A1S	10	0.718	6.9		10	If	10	69
EKT-fuel gas heating	90A1S	4	0.337	1.1		350	If	350	379
EKT-fuel gas heating	90A1S	4	0.337	1.0		250	If	250	250
EKT-fuel gas heating	90A1S	4	0.337	1.8		20	If	20	35
EKT-fuel gas heating	90A1S	1.5	0.200	1.0		12	If	12	12
EKT-fuel gas heating	90A1S	1	0.179	1.0		10	If	10	10
Estimate GAC-raw water						**25**	**If**		**126**
GAC-raw water	15A1S	10	std wt	5.8		20	If	20	116
GAC-raw water	15A1S	1.5	xs	1.9		3	If	3	6
GAC-raw water	15A1S	1	xs	2.1		2	If	2	4
Estimate GHC-condensate makeup						**645**	**If**		**761**
GHC-condensate makeup	15D1S	6	10s	5.5		10	If	10	55
GHC-condensate makeup	15D1S	4	10s	1.9		75	If	75	141
GHC-condensate makeup	15D1S	3	10s	1.0		385	If	385	375
GHC-condensate makeup	15D1S	2	40s	1.0		150	If	150	155
GHC-condensate makeup	15D1S	1	40s	1.4		25	If	25	36
Estimate GKB-potable water						**875**	**If**		**1016**
GKB-potable water	15M61S	3	std wt	2.7		25	If	25	67
GKB-potable water	15M61S	2	xs	1.0		250	If	250	249
GKB-Potable water	15M61S	1.5	xs	1.1		500	If	500	569
GKB-potable water	15M61S	1	xs	1.3		100	If	100	131
Estimate GMA-wastewater collection						**1370**	**If**		**1479**
GMA-wastewater collection	15D11S	1	40s	2.1		10	If	10	21
GMA-wastewater collection	15D11S	1.5	40s	2.1		10	If	10	21
GMA-wastewater collection	15A2S	4	std wt	1.6		100	If	100	161
GMA-wastewater collection	60D3S	1	40s	2.1		10	If	10	21
GMA-wastewater collection	15D11S	1.5	40s	2.1		10	If	10	21
GMA-wastewater collection	60D1S	4	10s	3.3		100	If	100	332
GMA-wastewater collection	15A1S	8	std wt	4.4		40	If	40	175
GMA-wastewater collection	15A1S	4	std wt	2.0		60	If	60	119
GMA-wastewater collection	15A1S	2	xs	0.5		1000	If	1000	533
GMA-wastewater collection	15A1S	1.5	xs	2.0		20	If	20	40
GMA-wastewater collection	15A1S	1	xs	3.5		10	If	10	35

6.27.2 Aboveground piping sheet 2

Line no.	Specification			MH	Qty	Unit	Qty	PF
	Material	Size	Sch/Thk	MH_a	n_i		n_i	$n_i MH_a$
Estimate HAN-boiler blowdown					4033	lf		5484
HAN-boiler blowdown	15A1S	1	xs	6.7	5	lf	5	34
HAN-boiler blowdown	15A1S	1.5	xs	3.6	8	lf	8	29
HAN-boiler blowdown	15A1S	2	xs	1.6	20	lf	20	32
HAN-boiler blowdown	15A1S	3	std wt	3.6	10	lf	10	36
HAN-boiler blowdown	15A1S	4	std wt	4.6	20	lf	20	92
HAN-boiler blowdown	15A5S	2	xs	1.7	10	lf	10	17
HAN-boiler blowdown	15A5S	6	std wt	2.3	10	lf	10	23
HAN-boiler blowdown	15A5S	24	std wt	3.2	70	lf	70	221
HAN-boiler blowdown	15A5S	1.5	xs	0.9	1200	lf	1200	1066
HAN-boiler blowdown	15C3S	1.5	xs-P22	4.9	40	lf	40	194
HAN-boiler blowdown	15C3S	12	std wt-P22	6.7	10	lf	10	67
HAN-boiler blowdown	60C1S	1.5	xs-P91	1.3	1700	lf	1700	2133
HAN-boiler blowdown	60C5S	12	std wt-P91	3.4	20	lf	20	67
HAN-boiler blowdown	250C1S	1.5	xs-P91	1.6	900	lf	900	1441
HAN-boiler blowdown	250C1S	12	std wt-P91	3.4	10	lf	10	34
Estimate HSJ-anhydrous ammonia					354	lf		238
HSJ-anhydrous ammonia	30D1S	2	40s	0.6	350	lf	350	224
HSJ-anhydrous ammonia	30D1S	1.5	40s	3.3	4	lf	4	13
Estimate LAB-boiler feedwater					1344	lf		2273
LAB-boiler feedwater	15A1S	2	xs	1.0	130	lf	130	128
LAB-boiler feedwater	30A1S	10	std wt	3.4	120	lf	120	412
LAB-boiler feedwater	30A1S	6	std wt	2.2	100	lf	100	215
LAB-boiler feedwater	30A1S	4	std wt	1.3	70	lf	70	89
LAB-boiler feedwater	30A1S	1.5	xs	4.8	8	lf	8	39
LAB-boiler feedwater	30A1S	1.5	xs	4.0	11	lf	11	44
LAB-boiler feedwater	90A1S	6	xs	2.2	165	lf	165	366
LAB-boiler feedwater	90A1S	4	xs	1.7	70	lf	70	121
LAB-boiler feedwater	90A1S	2	xs	0.9	180	lf	180	160
LAB-boiler feedwater	90A1S	1.5	xs	2.7	14	lf	14	38
LAB-boiler feedwater	90A1S	1	xs	4.7	4	lf	4	19
LAB-boiler feedwater	250A1S	8	.84″ CS	2.5	160	lf	160	396
LAB-boiler feedwater	250A1S	2	160	0.9	196	lf	196	182
LAB-boiler feedwater	250A1S	1.5	160	2.4	18	lf	18	44
LAB-boiler feedwater	90A1S	1	xs	4.7	4	lf	4	19

6.27.3 Aboveground piping sheet 3

Line no.	Specification			MH	Qty	Unit	Qty	PF
	Material	Size	Sch/Thk	MH_a	n_i		n_i	$n_i MH_a$
Estimate LBA-high-pressure steam					530	lf		2484
LBA-high-pressure steam	250C1S	14	160 P91	46.8	2	lf	2	94
LBA-high-pressure steam	250C1S	10	160 P91	5.7	270	lf	270	1531
LBA-high-pressure steam	250C1S	6	160 P91	12.3	9	lf	9	110
LBA-high-pressure steam	250C1S	3	160 P91	4.0	10	lf	10	40

(Continued)

(Continued)

	Specification			MH	Qty	Unit	Qty	PF
Line no.	Material	Size	Sch/Thk	MH$_a$	n$_i$		n$_i$	n$_i$ MH$_a$
LBA-high-pressure steam	250C1S	2	xs-P91	1.0	70	lf	70	69
LBA-high-pressure steam	250C1S	2	xs-P91	1.3	45	lf	45	59
LBA-high-pressure steam	250C1S	1.5	xs-P91	2.4	10	lf	10	24
LBA-high-pressure steam	250C1S	1	xs-P91	2.1	16	lf	16	34
LBA-high-pressure steam	250C1S	10	160 P91	20.2	12	lf	12	242
LBA-high-pressure steam	60C5S	14	std wt-P91	2.6	30	lf	30	78
LBA-high-pressure steam	60A1S	14	40	5.7	15	lf	15	86
LBA-high-pressure steam	250C1S	2	xs-P91	3.8	15	lf	15	57
LBA-high-pressure steam	250C1S	1.5	xs-P91	2.7	6	lf	6	16
LBA-high-pressure steam	250C1S	1	xs-P91	2.1	16	lf	16	34
LBA-high-pressure steam	60C5S	2	xs-P22	2.7	4	lf	4	11
Estimate LBB-hot reheat steam					**473**	**lf**		**2269**
LBB-hot reheat steam	60CS1	20	xs-P91	4.2	260	lf	260	1100
LBB-hot reheat steam	60CS1	16	std wt-P91	5.1	30	lf	30	153
LBB-hot reheat steam	60CS1	10	std wt-P91	5.6	6	lf	6	34
LBB-hot reheat steam	60CS1	8	std wt-P91	4.5	6	lf	6	27
LBB-hot reheat steam	60CS1	2	xs-P91	2.5	60	lf	60	153
LBB-hot reheat steam	60CS1	1.5	xs-P91	1.6	12	lf	12	19
LBB-hot reheat steam	60CS1	1	xs-P91	1.5	10	lf	10	15
LBB-hot reheat steam	60CS1	20	xs-P91	11.5	15	lf	15	172
LBB-hot reheat steam	60CS1	30	xs-P91	8.5	70	lf	70	592
LBB-hot reheat steam	60CS1	1	xs-P91	1.5	4	lf	4	6

6.27.4 Aboveground piping sheet 4

	Specification			MH	Qty	Unit	Qty	PF
Line no.	Material	Size	Sch/Thk	MH$_a$	n$_i$		n$_i$	n$_i$ MH$_a$
Estimate LBC-cold reheat steam					**505**	**lf**		**1950**
LBC-cold reheat steam	60A1S	24	60	32.0	2	lf	2	64
LBC-cold reheat steam	60A1S	20	40	6.4	110	lf	110	704
LBC-cold reheat steam	60C5S	20	xs-P91	6.0	110	lf	110	665
LBC-cold reheat steam	60C5S	4	std wt-P91	1.3	25	lf	25	34
LBC-cold reheat steam	60C5S	4	std wt-P91	1.0	100	lf	100	98
LBC-cold reheat steam	60C5S	3	std wt-P91	1.8	20	lf	20	37
LBC-cold reheat steam	60C5S	10	std wt-P91	15.7	6	lf	6	94
LBC-cold reheat steam	60C5S	2	xs-P91	2.7	45	lf	45	121
LBC-cold reheat steam	60C5S	1.5	xs-P91	1.7	4	lf	4	7
LBC-cold reheat steam	60C5S	1	xs-P91	1.4	8	lf	8	11
LBC-cold reheat steam	90A1S	2	xs	1.0	20	lf	20	21
LBC-cold reheat steam	90A1S	1.5	xs	1.2	4	lf	4	5
LBC-cold reheat steam	90A1S	1	xs	1.2	4	lf	4	5
LBC-cold reheat steam	60C5S	2	xs-P22	3.5	3	lf	3	11
LBC-cold reheat steam	60A1S	6	std wt	1.9	20	lf	20	37
LBC-cold reheat steam	60A1S	10	std wt	4.4	3	lf	3	13
LBC-cold reheat steam	60A1S	2	xs	1.1	15	lf	15	16
LBC-cold reheat steam	60A1S	1.5	xs	1.3	4	lf	4	5
LBC-cold reheat steam	60A1S	1	xs	1.3	2	lf	2	3

(Continued)

(Continued)

Line no.	Material	Size	Sch/Thk	MH MH$_a$	Qty n$_i$	Unit	Qty n$_i$	PF n$_i$ MH$_a$
Estimate LBG-auxiliary steam					**490**	**lf**		**669**
LBG-auxiliary steam	30A5S	6	std wt	2.1	120	lf	120	248
LBG-auxiliary steam	30A5S	4	std wt	1.1	200	lf	200	228
LBG-auxiliary steam	30A5S	4	std wt	1.3	100	lf	100	130
LBG-auxiliary steam	30A5S	3	std wt	0.9	70	lf	70	63
Estimate-LBM-low-pressure steam					**462**	**lf**		**1229**
LBM-low-pressure steam	15A5S	14	std wt	2.7	200	lf	200	543
LBM-low-pressure steam	15A5S	10	std wt	2.4	30	lf	30	73
LBM-low-pressure steam	15A5S	2	xs	1.9	50	lf	50	94
LBM-low-pressure steam	15A5S	8	std wt	12.8	3	lf	3	39
LBM-low-pressure steam	15A5S	6	std wt	11.7	3	lf	3	35
LBM-low-pressure steam	15A5S	2	xs	0.9	60	lf	60	52
LBM-low-pressure steam	15A5S	12	std wt	4.5	15	lf	15	67
LBM-low-pressure steam	15C3S	18	std wt-P22	5.0	45	lf	45	225
LBM-low-pressure steam	15A5S	1.5	xs	1.3	2	lf	2	3
LBM-low-pressure steam	15C3S	1	xs-P22	1.6	2	lf	2	3
LBM-low-pressure steam	30A1S	2	xs	2.3	30	lf	30	68
LBM-low-pressure steam	30A1S	1.5	xs	1.2	6	lf	6	7
LBM-low-pressure steam	30A1S	1	xs	1.2	8	lf	8	9
LBM-low-pressure steam	30A5S	2	xs	1.2	8	lf	8	10

6.27.5 Aboveground piping sheet 5

Line no.	Material	Size	Sch/Thk	MH MH$_a$	Qty n$_i$	Unit	Qty n$_i$	PF n$_i$ MH$_a$
Estimate LCA-condensate					**1626**	**lf**		**3154**
LCA-condensate	15A1S	18	std wt	11.8	50	lf	50	592
LCA-condensate	15A1S	6	std wt	4.4	10	lf	10	44
LCA-condensate	15A1S	3	std wt	2.0	10	lf	10	20
LCA-condensate	15A1S	2	xs	0.8	70	lf	70	54
LCA-condensate	15A1S	1.5	xs	2.5	10	lf	10	25
LCA-condensate	15A1S	1	xs	2.1	40	lf	40	86
LCA-condensate	15A1S	0.8	xs	2.0	2	lf	2	4
LCA-condensate	30A1S	10	std wt	2.3	350	lf	350	794
LCA-condensate	30A1S	10	std wt	2.9	110	lf	110	319
LCA-condensate	30A1S	6	std wt	2.2	100	lf	100	215
LCA-condensate	30A1S	4	std wt	3.0	30	lf	30	89
LCA-condensate	30A1S	3	std wt	1.3	205	lf	205	260
LCA-condensate	30A1S	2	xs	0.8	130	lf	130	100
LCA-condensate	30A1S	1.5	xs	1.0	425	lf	425	414
LCA-condensate	30A1S	1	xs	1.7	15	lf	15	25
LCA-condensate	30A1S	0.8	xs	2.1	1	lf	1	2
LCA-condensate	90A1S	6	xs	1.7	60	lf	60	101
LCA-condensate	90A1S	1.5	xs	1.2	4	lf	4	5
LCA-condensate	90A1S	1	xs	1.2	4	lf	4	5
Estimate LDK-condensate polishing					**370**	**lf**		**732**
LDK-condensate polishing	15A1S	1.5	xs	0.8	200	lf	200	170

(*Continued*)

(Continued)

	Specification			MH	Qty	Unit	Qty	PF
Line no.	Material	Size	Sch/Thk	MH_a	n_i		n_i	$n_i MH_a$
LDK-condensate polishing	15A1S	10	std wt	3.8	60	lf	60	229
LDK-condensate polishing	15A1S	12	std wt	3.0	110	lf	110	332
Estimate MA-steam turbine					**500**	**lf**		**593**
MA-steam turbine	15A1S	1	xs	2.0	30	lf	30	60
MA-steam turbine	15A1S	1.5	xs	0.8	150	lf	150	121
MA-steam turbine	15A1S	2	xs	0.8	200	lf	200	160
MA-steam turbine	15A1S	4	std wt	2.3	30	lf	30	68
MA-steam turbine	15A5S	4	std wt	1.0	30	lf	30	30
MA-steam turbine	15A5S	6	std wt	1.3	40	lf	40	51
MA-steam turbine	15D1S	6	10s	5.1	20	lf	20	103
Estimate MAJ-air extraction					**320**	**lf**		**537**
MAJ-air extraction	15A1S	24	std wt	4.1	40	lf	40	162
MAJ-air extraction	15A1S	8	std wt	1.5	190	lf	190	279
MAJ-air extraction	15A1S	6	std wt	3.0	6	lf	6	18
MAJ-air extraction	15A1S	6	std wt	6.1	4	lf	4	25
MAJ-air extraction	15A1S	0.8	xs	0.7	80	lf	80	54

6.27.6 Aboveground piping sheet 6

	Specification			MH	Qty	Unit	Qty	PF
Line no.	Material	Size	Sch/Thk	MH_a	n_i		n_i	$n_i MH_a$
Estimate MAL-steam turbine drains					**2900**	**lf**		**2973**
MAL-steam turbine drains	15A5S	1.5	xs	0.9	490	lf	490	437
MAL-steam turbine drains	15A5S	2	xs	1.0	275	lf	275	283
MAL-steam turbine drains	15C3S	1.5	xs-P22	1.1	70	lf	70	74
MAL-steam turbine drains	250C1S	1.5	xs-P91	1.0	400	lf	400	389
MAL-steam turbine drains	250C1S	2	xs-P91	1.1	250	lf	250	281
MAL-steam turbine drains	250C1S	3	160 P91	1.3	65	lf	65	85
MAL-steam turbine drains	30A5S	2	xs	0.9	380	lf	380	330
MAL-steam turbine drains	60C1S	1.5	xs-P91	1.1	300	lf	300	317
MAL-steam turbine drains	60C1S	2	xs-P91	1.1	250	lf	250	277
MAL-steam turbine drains	60C5S	2	xs-P22	1.0	220	lf	220	222
MAL-steam turbine drains	60C5S	4	std wt-P91	1.0	100	lf	100	98
MAL-steam turbine drains	316SS	6	40s	1.8	100	lf	100	178
Estimate MB-combustion turbine					**508**	**lf**		**719**
MB-combustion turbine	15A1S	1	xs	1.8	8	lf	8	14
MB-combustion turbine	15A1S	1.5	xs	0.9	80	lf	80	73
MB-combustion turbine	15A1S	2	xs	1.1	40	lf	40	44
MB-combustion turbine	15A1S	6	std wt	3.2	12	lf	12	38
MB-combustion turbine	15A1S	6	std wt	6.3	8	lf	8	50
MB-combustion turbine	15D1S	1	40s	2.8	10	lf	10	28
MB-combustion turbine	15D1S	1.5	40s	1.1	100	lf	100	108
MB-combustion turbine	15D1S	2	40s	1.1	60	lf	60	69
MB-combustion turbine	15D1S	6	10s	2.1	80	lf	80	170
MB-combustion turbine	30A3S	1.5	xs	1.1	40	lf	40	43

(Continued)

(Continued)

Line no.	Specification			MH	Qty	Unit	Qty	PF
	Material	Size	Sch/Thk	MH_a	n_i		n_i	$n_i\,MH_a$
MB-combustion turbine	30D1S	2	40s	1.3	30	lf	30	39
MB-combustion turbine	60A4S	1.5	xs	1.1	40	lf	40	43
Estimate PAB-circulating water					**710**	**lf**		**7759**
PAB-circulating water	15A2S	54	.500″ Wall	20.9	70	lf	70	1465
PAB-circulating water	15A2S	42	.375″ Wall	9.7	165	lf	165	1593
PAB-circulating water	15A2S	36	.375″ Wall	6.8	140	lf	140	953
PAB-circulating water	15A2S	30	.375″ Wall	23.4	35	lf	35	818
PAB-circulating water	15A2S	24	std wt	9.3	60	lf	60	556
PAB-circulating water	15A2S	18	std wt	18.3	50	lf	50	915
PAB-circulating water	15A2S	16	std wt	12.9	50	lf	50	645
PAB-circulating water	15A2S	6	std wt	9.9	20	lf	20	197
PAB-circulating water	15A2S	4	std wt	5.5	40	lf	40	221
PAB-circulating water	15D11S	1	40s-Duplex ss	6.2	40	lf	40	249
PAB-circulating water	15D11S	1.5	40s-Duplex ss	3.7	40	lf	40	147

6.27.7 Aboveground piping sheet 7

Line no.	Specification			MH	Qty	Unit	Qty	PF
	Material	Size	Sch/Thk	MH_a	n_i		n_i	$n_i\,MH_a$
Estimate PBN-circulating water chemical feed					**1225**	**lf**		**1242**
PBN-circulating water chemical feed	12J4S	2	80-CPVC	1.0	40	lf	40	40
PBN-circulating water chemical feed	12J4S	1	80-CPVC	1.0	550	lf	550	550
PBN-circulating water chemical feed	15D2S	2	80s-Alloy 20	1.1	45	lf	45	50
PBN-circulating water chemical feed	15D2S	0.5	80s-Alloy 20	1.0	40	lf	40	40
PBN-circulating water chemical feed	15D11S	1.5	40s-Duplex ss	1.0	70	lf	70	70
PBN-circulating water chemical feed	15D1S	1	40s	1.1	40	lf	40	44
PBN-circulating water chemical feed	150Q1S	0.5	ss tube	1.0	120	lf	120	120
PBN-circulating water chemical feed	15D1S	1	40s	1.1	40	lf	40	44
PBN-circulating water chemical feed	150Q1S	0.5	ss tube	1.0	120	lf	120	120
PBN-circulating water chemical feed	15D1S	1	40s	1.1	40	lf	40	44
PBN-circulating water chemical feed	150Q1S	0.5	ss tube	1.0	120	lf	120	120
Estimate PCB-service water					**1620**	**lf**		**2271**
PCB-service water	15A1S	14	std wt	6.9	50	lf	50	345
PCB-service water	15A1S	10	std wt	3.9	40	lf	40	154
PCB-service water	15A1S	6	std wt	3.3	70	lf	70	232
PCB-service water	15A1S	4	std wt	1.0	300	lf	300	305
PCB-service water	15A1S	3	std wt	1.0	205	lf	205	211
PCB-service water	15A1S	2	xs	1.0	275	lf	275	275
PCB-service water	15A1S	1.5	xs	1.1	680	lf	680	748
Estimate PGB-closed cooling water					**3135**	**lf**		**6548**
PGB-closed cooling water	15A1S	24	std wt	6.9	130	lf	130	892
PGB-closed cooling water	15A1S	16	std wt	4.9	90	lf	90	443
PGB-closed cooling water	15A1S	14	std wt	2.3	530	lf	530	1208

(*Continued*)

(Continued)

Line no.	Specification			MH	Qty	Unit	Qty	PF
	Material	Size	Sch/Thk	MH_a	n_i		n_i	$n_i\,MH_a$
PGB-closed cooling water	15A1S	10	std wt	2.3	420	lf	420	947
PGB-closed cooling water	15A1S	8	std wt	1.6	520	lf	520	815
PGB-closed cooling water	15A1S	6	std wt	2.2	410	lf	410	888
PGB-closed cooling water	15A1S	4	std wt	1.3	250	lf	250	334
PGB-closed cooling water	15A1S	3	std wt	1.5	160	lf	160	238
PGB-closed cooling water	15A1S	2	xs	1.1	95	lf	95	105
PGB-closed cooling water	15A1S	1.5	xs	1.1	480	lf	480	533
PGB-closed cooling water	15A1S	1	xs	2.9	50	lf	50	146

6.27.8 Aboveground piping sheet 8

Line no.	Specification			MH	Qty	Unit	Qty	PF
	Material	Size	Sch/Thk	MH_a	n_i		n_i	$n_i\,MH_a$
Estimate QC-cycle chemical feed					880	lf		880
QC-cycle chemical feed	316 SS	0.5	80s-316 ss	1.0	350	lf	350	350
QC-cycle chemical feed	316 SS	0.5	80s-316 ss	1.0	350	lf	350	350
QC-cycle chemical feed	316 SS	0.5	80s-316 ss	1.0	180	lf	180	180
Estimate QFB-instrument/ service air					5600	lf		6447
QFB-instrument/service air	15D7S	3	5S-304SS Pressfit	1.0	600	lf	600	613
QFB-instrument/service air	15D7S	2	5S-304SS Pressfit	1.2	1200	lf	1200	1446
QFB-instrument/service air	15D7S	1.5	5S-304SS Pressfit	1.3	1000	lf	1000	1349
QFB-instrument/service air	15D7S	1	5S-304SS Pressfit	1.2	1000	lf	1000	1239
QFB-instrument/service air	150Q1S	0.5	80s-316 ss	1.0	1800	lf	1800	1800
Estimate QH-aux boiler					650	lf		1057
QH-aux boiler	15A1S	1	xs	2.0	10	lf	10	20
QH-aux boiler	15A1S	1.5	xs	1.0	100	lf	100	100
QH-aux boiler	15A1S	2	xs	1.1	100	lf	100	108
QH-aux boiler	15A1S	4	std wt	2.0	40	lf	40	82
QH-aux boiler	15A5S	6	std wt	2.5	50	lf	50	124
QH-aux boiler	15A5S	4	std wt	2.1	30	lf	30	63
QH-aux boiler	15D1S	1	40s	2.8	10	lf	10	28
QH-aux boiler	15D1S	1.5	40s	1.2	50	lf	50	62
QH-aux boiler	15D1S	2	40s	1.2	50	lf	50	62
QH-aux boiler	30A5S	6	std wt	5.8	30	lf	30	175
QH-aux boiler	30A5S	4	std wt	1.4	45	lf	45	65
QH-aux boiler	30A5S	3	std wt	2.7	15	lf	15	40
QH-aux boiler	30A5S	2	xs	1.0	115	lf	115	115
QH-aux boiler	30A5S	1	xs	2.6	5	lf	5	13
Estimate QJ-nitrogen	15A1S	1	xs	1.1	**500**	**lf**	500	550
Estimate QU-sampling and analysis					5442	lf		18,884
QU-sampling and analysis	150Q1S	0.25	ss tube	2.0	3500	lf	3500	7000

(Continued)

(Continued)

Line no.	Specification			MH	Qty	Unit	Qty	PF
	Material	Size	Sch/Thk	MH_a	n_i		n_i	$n_i MH_a$
QU-sampling and analysis	150Q1S	0.375	ss tube	2.0	1100	lf	1100	2200
QU-sampling and analysis	150Q1S	0.5	ss tube	2.0	17	lf	17	34
QU-sampling and analysis	150Q1S	0.75	ss tube	2.0	17	lf	17	34
QU-sampling and analysis	250Q2S	0.25	ss tube	2.0	200	lf	200	400
QU-sampling and analysis	250Q2S	0.5	ss tube	2.0	1	lf	1	2
QU-sampling and analysis	250Q2S	0.75	ss tube	2.0	1	lf	1	2
QU-sampling and analysis	250Q4S	0.75	ss tube	2.0	200	lf	200	400
QU-sampling and analysis	250Q4S	0.25	ss tube	22.0	400	lf	400	8800
QU-sampling and analysis	250Q4S	0.5	ss tube	2.0	3	lf	3	6
QU-sampling and analysis	250Q4S	0.75	ss tube	2.0	3	lf	3	6

6.28 Heat recovery steam generator configuration

There are many different configurations for combined cycle power plants:

- Each gas turbine has one associated HRSG.
- 2 × 1 Configuration, two GT/HRSG trains supply one STG.
- Multiple HRSG trains supply steam to one STG.
- There can be 1 × 1, 3 × 1, and 4 × 1 arrangements.

6.29 Combined cycle power plant equipment man-hour breakdown

The direct craft man-hours, provided in this manual, have been cost coded, collected in the field, summarized, and verified by statistical analysis. Man-hours are for direct labor and do not include indirect and staff labor. The craft man-hours can be modified as the user's experience and situation require.

Combined cycle power plant
Owner-furnished equipment

6.29.1 Combined cycle power plant: configuration: 2 × 1

Direct craft man-hours	Actual	Qty	BM	IW	MW	Estimate
Scope of work	MH		MH	MH	MH	MH
E and F class CTG-equipment installation man-hours	13,260	2	14,462	7674	4385	26,521
Reheat double-flow STG-installation man-hours	10,420	1	1636	1335	7449	10,420
	24,258	2	48,516			48,516

(*Continued*)

(Continued)

Direct craft man-hours	Actual	Qty	BM	IW	MW	Estimate
Scope of work	MH		MH	MH	MH	MH
HRSG triple pressure; three wide-installation man-hours Air cooled condenser (36 cell)-installation man-hours	89,604	1	40,176	33,948	15,480	89,604
Balance of plant installation man-hours	6644	1	1475	1552	3618	6644

6.29.2 Combined cycle power plant owner-furnished equipment

Direct craft man-hours	Actual	BM	IW	MW		
Scope of work	MH	MH	MH	MH	MH	Cells
E and F class CTG-equipment installation man-hours	13,260	7231	3837	2192	13,260	
Reheat double-flow STG-installation man-hours	10,420	1636	1335	7449	10,420	
HRSG triple pressure; three wide-installation man-hours	24,258	24,258			24,258	
HRSG triple pressure; double wide-installation man-hours	16,500	16,500			16,500	
HRSG double pressure; single wide-installation man-hours	7456	7456			7456	
HRSG single pressure; double wide-installation man-hours	30,958	30,958			30,958	
Air cooled condenser (36 cell)-installation man-hours	89,605	40,176	33,948	15,480	89,604	36
Surface condenser-equipment installation man-hours	3460	3460			3460	
Balance of plant installation man-hours	6644	1475	1552	3618	6644	
ZLD-equipment installation man-hours	6969		6156	840	6996	

6.30 Direct craft man-hour summary

6.30.1 Combined cycle power plant: configuration: 1 × 1

Direct craft man-hours	Estimate					
Scope of work	MH	Carpenter	Labor	IW	MW	PF
Foundations						
Boiler foundations	7347	1138	4550	1659		
CTG and CTG-related foundations	5528	1025	3330	1173		
STG and STG-related foundations	15,865	2939	9471	3455		
Plant mechanical system–related foundations	7161	2046	3962	1152		
Utility distribution plant electrical controls foundations	17,628	5932	8138	3557		
Mechanical equipment (CTG, STG, HRSG), w/piping			BM	IW	MW	PF
E and F class CTG-equipment installation man-hours	13,260		7231	3837	2192	
Reheat double-flow STG-installation man-hours	10,420		1636	1335	7449	
HRSG triple pressure; Three wide-installation man-hours	24,258		24,258			
Balance of plant installation man-hours	6644		1475	1552	3618	
HRSG—large bore code piping	15,529					15,529
HRSG—small bore code piping	5191					5191
HRSG—risers and down comers	3827					3827
HRSG—field trim piping	3341					3341
SP-01 AIG piping	2262					2262
STG—vendor piping	7014					7014
STG piping	5928					5928

(Continued)

(Continued)

Direct craft man-hours	Estimate					
Scope of work	MH	Carpenter	Labor	IW	MW	PF
CTG—vendor piping	14,634					14,634
BOP equipment		BM	IW	MW		
Balance of plant installation estimate	6644	1475	1552	3618		
Structural steel				IW		
Structural steel installation estimate	11,837			11,837		
Underground piping						PF
Underground piping	13,913					13,913
Aboveground piping						PF
Aboveground piping	77,741					77,741
Direct craft man-hours	**273,712**	**14,556**	**65,603**	**33,176**	**13,259**	**147,118**

Chapter 7

Gasifier labor estimate

7.1 Introduction

This chapter provides the readers a basic understanding of the fundamentals and the operating relationship between the plant equipment in the gasification process and covers the craft labor for the assembly and field erection required to put the equipment into operation in a gasification plant.

Gasification is a process that converts organic- or fossil fuel−based carbonaceous materials into carbon monoxide, hydrogen, and carbon dioxide. This is achieved by reacting the material at high temperatures without combustion, with a controlled amount of oxygen and/or steam. The resulting gas mixture is called syngas or producer gas and is itself a fuel. The power derived from gasification and combustion of the resultant gas is considered to be a source of renewable energy if the gasified compounds were obtained from biomass. Gasification is a technology that converts carbon-containing materials, including coal, waste, and biomass, into synthetic gas that in turn can be used to produce electricity and other valuable products, such as chemicals, fuels, and fertilizers.

Industrial Construction Estimating Manual. DOI: https://doi.org/10.1016/B978-0-12-823362-7.00007-7

7.2 Gasifier bid breakdown

7.2.1 Total direct craft man-hours sheet 1

	BM-MH
7.2 Gasifier Bid Breakdown	
7.2.1 Total Direct Craft Man-Hours Sheet 1	**95124**
Char Distribution Hopper A, B, C (56'x25'x25')	1904
Pulverized Feedstock Distribution Hopper A, B, C (48'x21'x21')	2374
PF Outlet Bypass Shutoff Valve*** Gate Valve (13.9'x3.1'x4.3')	120
Pulverized Feedstock Drying Gas Duct (High Temp.)	4593
Pulverized Feedstock Drying Gas Duct (Low Temp.)	2793
Char Cyclone (68'x20'x20')	946
Pulverized Feedstock Collector A, B (40.3'x27.3'x37.4')	520
Pulverized Feedstock Drying Gas Duct (from Pulverized Feedstock Collector to Stack)	7505
Feedstock Gate A, B Motorized Slide Gate (5.9'x7.5'x4.2')	240
Feedstock Bunker A, B (40'D x 101')	3360
PF Blowback N2 Heater (13.2'x3.2'x4.6')	60
N2 Heater (16.7'x3.9'x6.4)	80
High Pressure Soot Blower	98
Porous Filter A, B (39'x19'x18')	1819
Porous Filter Element for PF-A, B	9760
PF Blowback N2 Buffer Tank (38.4'x8.8'x8.8)	240
Pulverized Feedstock Distribution Hopper Exhaust Collector (22.9'x16.4'x36.1')	100
High Pressure Soot Blower	294
Gasifier and SGC Pressure Vessel (GP1) (62'x23'x22')	240
Gasifier and SGC Pressure Vessel (GP2) (60'x23'x21')	4120
Gasifier and SGC Pressure Vessel (GP3) (49'x22'x23')	4120
Gasifier and SGC Pressure Vessel (GP4) (57'x23'x23')	4360
GP1.GP2, GP3, GP4	650
Gasifier and SGC Water Wall, Economizer, 1ry-SH (GW1); Membrane Wall 1.5" Día; Econ Tube 1.77" Día	1924
Gasifier and SGC Water Wall, 1ry-SH, 2ry-SH, Evaporator (GW2); Membrane Wall 1.5" Día; Eva Tube 1.77" Día	2102
Gasifier and SGC Water Wall, Evaporator (GW3); Membrane Wall 1.5" Día; Eva Tube 1.77" Día	1924
Gasifier and SGC Water Wall (GW4); Membrane Wall 1.5" Día	1350
Gasifier and SGC Water Wall (GW5); Membrane Wall 1.5" Día	1350
GW1, GW2, GW3, GW4 & GW5	4070
Gasifier Start-up Burner A, B (29.2'x4.9'x4.3')	120
Gasification Solids Melting Burner A.B (28.5'x8.5'x6.1')	120
Char Burner A, B, C, D (21.3'x4.6'x4.6')	240
Pulverized Feedstock Distributor for Redactor Burner (4'x4'x4')	20
Pulverized Feedstock Distributor for Combustor Burner (4'x4'x4')	20
Mechanical Equipment	9182
Feedstock Drying Gas Stack (16.8' Diameterx8.4'x8.3')	1656
Gasification Solids Discharge Conveyor (74.3'x11.4'x26.2')	1280
Gasifier Circulation Pump Cooler A, B & Water Cooler	70
Feedstock Conveyer Room Ventilation Fan A, B	60
Feedstock Pulverize & Pulverized Feedstock Feed System	6170
Pulverized Feedstock Drying System	3950
Pulverized Feedstock Burner	1520
Char Recovery and Feed System	1720
Gasification Solids Disposal System	5980

7.3 Detailed estimate using the unit quantity model to erect gasifier

7.3.1 Equipment estimate sheet 1

7.3.1 Equipment Estimate Sheet 1	Qty	Qty	Qty			
Description	Wt. (ton)	Hdl/Set	Field Jt	Bolt Up	MH/Qty	BM-MH
Char Distribution Hopper						
Char Distribution Hopper A, B, C (56'x25'x25')	234	3				1904
Char Distribution Hopper A		By Others				
Char Distribution Hopper Fluidization Chamber A		By Others				
Char Distribution Hopper Fluidization Filter A		3			20.00	60
Char Distribution Hopper Fluidization Filter Retainer A		3		444	0.44	193
Steam Tracing		By Others				
Weld Stopper 1.26" Thk x 47.24" L BW			1134		1.35	1531
Weld Manhole Nozzle 26" Diameter		3	245		0.49	120
Pulverized Feedstock Distribution Hopper A, B, C (48'x21'x21')	131	3				2374
Pulverized Feedstock Distribution Hopper A		By Others				
Pulverized Feedstock Distribution Hopper Fluidization Chamber A		By Others				
Pulverized Feedstock Distribution Hopper Fluidization Filter A		3			20.00	60
Pulverized Feedstock Distribution Hopper Fluidization Filter Retainer A		3		312	0.49	154
Weld Stopper 1.26" Thk x 62.99" L BW			1512		1.35	2041
Weld Manhole Nozzle 26" Diameter		3	245		0.49	120
Steam Tracing		By Others				
PF Outlet Bypass Shutoff Valve* Gate Valve (13.9'x3.1'x4.3')**	6	1	2		60.00	120

7.3.2 Equipment estimate sheet 2

Description	Wt. (ton)	Hdl/Set	Field Jt	Bolt Up	MH/Qty	BM-MH
Pulverizer Drying Hot Gas Duct						
Pulverized Feedstock Drying Gas Duct (High Temp.)	82					4593
Receive and Transport Duct		21				420
Duct HB01 C 18.7'x11.9'x11.9'	6	1	142		0.39	55
Duct HB02 D 18.2'x15.6'x11.9'	8	1	285		0.32	91
Duct HB03 C 9.6'x11.9'x11.9'	3	1	142		0.39	55
Duct HB06 C 8.7'x11.9'x11.9'	3	1	142		0.39	55
Duct HB07 D 28.5'x15.5'x11.9'	11	1	443		0.30	131
Duct HB08 F 22.5'x14.5'x11.9'	8	1	594		0.28	169
Duct HB09 E 20.3'x11.9'x11.9'	5	1	142		0.39	55
Duct HB10 D 20.2'x17.1'x7.6'	6	1	345		0.31	106
Duct HB11 C 5.6'x9.8'x7.6'	1	1	74		0.52	39
Duct HB12 C 10.5'x9.8'x7.6'	2	1	74		0.52	39
Duct HB13 C 14.5'x9.8'x7.6'	3	1	74		0.52	39
Duct HB14 F 23.2'x8.8'x9.8'	5	1	432		0.30	128
Duct HB15 C 20.3'x9.8'x7.6'	4	1	74		0.52	39
Duct HB16 D 16.8'x8.9'x9.8'	3	1	150		0.38	57
Duct HB17 H 4.4'x8.7'x4.3'	1	1	19		1.31	25
Duct HB18 C 10.5'x9.8'x7.6'	2	1	74		0.52	39
Duct HB19 C 14.5'x9.8'x7.6'	3	1	74		0.52	39
Duct HB20 F 23.2'x8.8'x9.8'	5	1	432		0.30	128
Duct HB21 C 20.3'x9.8'x7.6'	4	1	74		0.52	39
Duct HB22 D 16.8x8.9'x7.6	3	1	150		0.38	57
Duct HB22 H 4.4'x8,9'x4.3	1	1	19		1.31	25
Duct Flange Weld			1646		0.35	576
Stiffener Corners Weld			2273		0.35	796
Corner Angle Weld			2812		0.35	984
Guide Vane Weld			107		0.35	37
Strut Weld			87		0.35	30
Accessory for Duct Weld			971		0.35	340

7.3.3 Equipment estimate sheet 3

| 7.3.3 Equipment Estimate Sheet 3 | | Qty | Qty | Qty | | |
Description	Wt. (ton)	Hdl/Set	Field Jt	Bolt Up	MH/Qty	BM-MH
Pulverizer Drying Hot/Cold Gas Duct						
Pulverized Feedstock Drying Gas Duct (Low Temp.)	**37**					**2793**
Receive and Transport Duct		17				340
Duct CB01 C	3		1	71	0.53	38
Duct CB02 D	3		1	184	0.36	66
Duct CB05 C	3		1	71	0.53	38
Duct CB06 C	1		1	71	0.53	38
Duct CB07 D	5		1	315	0.31	99
Duct CB08 F	4		1	600	0.28	170
Duct CB09 E	3		1	71	0.53	38
Duct CB10 D	3		1	293	0.32	93
Duct CB11 C	1		1	38	0.78	29
Duct CB12 C	1		1	38	0.78	29
Duct CB13 C	2		1	96	0.46	44
Duct CB14 D	2		1	56	0.60	34
Duct CB15 C	1		1	31	0.89	28
Duct CB16 C	1		1	70	0.54	37
Duct CB17 C	2		1	96	0.46	44
Duct CB10 D	2		1	56	0.60	34
Duct CB19 C	1		1	31	0.89	28
Flange Joint Weld				492	0.30	148
Stiffener Corners Weld				1955	0.35	684
Corner Angle Weld				1109	0.35	388
Guide Vane Weld				127	0.35	45
Strut Weld				61	0.35	21
Accessory for Duct Weld				687	0.35	240
Auxiliary Steam Drain Recovery Tank		1				40
Char Cyclone (68'x20'x20')	**286**	By Others				**946**
Steam Tracing		By Others				
Weld Stopper 1.26" Thk x 47.24" L BW				189	1.35	255
Weld Keeper Lug 0.98" Thk x 11.42" L BW				183	1.15	210
Base Plate Weld 0.63" Thk x 80.31" L FW				321	0.75	241
Weld Manhole Nozzle 26" Diameter		1		82	0.49	40
Syngas Pipe		By Others				
Char Pipe		By Others				
PF Outlet Safety Valve		By Others				
Char Cyclone Inlet Syngas Shutoff Valve*** Gate Valve		1	2		100.00	200
Pulverized Feedstock Pipe		By Others				
Pulverized Feedstock Collector A, B (40.3'x27.3'x37.4')	**195**	2			260.00	520

7.3.4 Equipment estimate sheet 4

7.3.4 Equipment Estimate Sheet 4		Wt. (ton)	Qty Hdl/Set	Qty Field Jt	Qty Bolt Up	MH/Qty	BM-MH
Description							
Pulverized Feedstock Drying Gas Duct (from Pulverized Feedstock Collector to Stack)		**125**					**7505**
Receive and Transport Duct			39				780
Duct DB01	C (7.2'x9.5'x5.6')	1		1		69	0.54	37
Duct DB02	D (12.9'x8.8'x9.5')	3		1		114	0.42	49
Duct DB03	C (10.5'x9.5'x5.6')	2		1		100	0.45	45
Duct DB04	C (17.7'x9.5'x5.6')	3		1		168	0.37	62
Duct DB05	D (14.9'x13.3'x5.6')	3		1		198	0.35	70
Duct DB06	C (16.4'x9.9'x5.6')	3		1		164	0.37	61
Duct DB07	E (13.3'x9.9'x9.9')	3		1		100	0.45	45
Duct DB08	C (7.2'x9.5'x5.6')	1		1		69	0.54	37
Duct DB09	D (12.9'x8.8'x9.5')	3		1		114	0.42	49
Duct DB10	C (10.5'x9.5'x5.6')	2		1		100	0.45	45
Duct DB11	C (17.7'x9.5'x5.6')	3		1		168	0.37	62
Duct DB12	F (13.5'x11.6'x9.9')	3		1		293	0.32	93
Duct DB13	C (15.7'x9.9'x9.9')	3		1		157	0.38	59
Duct DB14	C (17.1'x9.9'x9.9')	4		1		170	0.37	63
Duct DB15	D (15.4'x14.9'x9.9')	5		1		230	0.34	78
Duct DB16	C (7.2'x9.9'x9.9')	2		1		72	0.53	38
Duct DB17	D (16.1'x14.7'x9.9')	5		1		236	0.33	79
Duct DB18	C (15.1'x9.9'x9.9')	3		1		151	0.38	58
Duct DB19	C (17.1'x9.9'x9.9')	4		1		170	0.37	63
Duct DB20	C (9.8'x9.9'x9.9')	2		1		99	0.45	45
Duct DB21	F (23.2'x11.9'x9.9')	5		1		512	0.29	148
Duct DB22	D (19.0'x13.8'x9.9')	6		1		262	0.33	85
Duct DB23	C (13.8'x9.9'x9.9')	3		1		138	0.40	54
Duct DB24	G (15.7'x11.9'x10.2')	5		1		122	0.41	50
Duct DB51	E (7.8'x9.9'x9.9')	2		1		100	0.45	45
Duct DB52	D (20.5'x11.4'x9.9')	4		1		234	0.34	78
Duct DB53	C (20.3'x9.9'x9.9')	4		1		203	0.35	71
Duct DB54	C (20.3'x9.9'x9.9')	4		1		203	0.35	71
Duct DB55	C (20.3'x9.9'x9.9')	4		1		203	0.35	71
Duct DB56	C (25.6'x9.9'x9.9')	5		1		256	0.33	84
Duct DB57	C (20.7'x9.9'x9.9')	4		1		206	0.35	72
Duct DB58	C (20.7'x9.9'x9.9')	4		1		206	0.35	72
Duct DB59	C (5.6'x9.9'x9.9')	1		1		56	0.61	34
Duct DB60	C (10.5'x9.9'x9.9')	2		1		105	0.44	46
Duct DB61	D (14.9'x34.9'x9.9')	4		1		214	0.34	74
Duct DB62	D (14.9'x12.1'x9.9')	4		1		180	0.36	65
Duct DB63	C (13.8'x9.9'x9.9')	3		1		138	0.40	54
Duct DB64	C (17.1'x9.9'x9.9')	3		1		170	0.37	63
Duct DB65	E (12.8'x13.9'x9.9')	3		1		140	0.39	55
Flange Joint Weld				2338		0.30	701
Stiffener Corners Weld				4869		0.35	1704
Corner Angle Weld				3143		0.35	1100
Guide Vane Weld				324		0.35	113
Strut Weld				298		0.35	104
Accessory for Duct Weld				1639		0.35	574

7.3.5 Equipment estimate sheet 5

7.3.5 Equipment Estimate Sheet 5 Description	Wt. (ton)	Qty Hdl/Set	Qty Field Jt	Qty Bolt Up	MH/Qty	BM-MH
Feedstock Gate A, B Motorized Slide Gate (5.9'x7.5'x4.2')	2	2		4	60.00	240
Feedstock Bunker A, B (40'D x 101')	231	2			1680.00	3360
PF Blowback N2 Heater (13.2'x3.2'x4.6')	4	1			60.00	60
N2 Heater (16.7'x3.9'x6.4)	7	1			80.00	80
High Pressure Soot Blower	2					98
Handle Soot Blowers		2			10.00	20
High Pressure Soot Blower A, B (30'x3'x4")	2	2			24.00	48
High Pressure Shutoff Valve 6" Diameter Weld				2	15.00	30
Porous Filter A, B (39'x19'x18')	166					1819
Steam Tracing		By Others				
Weld Stopper 1.26" Thk x 62.99" L BW			126		1.75	220
Weld Keeper Lug 0.98" Thk x 11.42" L BW			504		1.35	680
Base Plate Weld 0.63" Thk x 80.31" L FW			365		1.15	420
Back Flow Pipe			642		0.75	482
Manhole Cover		2			8.00	16
Porous Filter Element for PF-A, B	12					9760
Scaffold & set rigging in vessel		2			400.00	800
Transport Elements to field site		2			160.00	320
Unpack the containers and transport element to MH		2			864.00	1728
Carry element in to vessel		2			864.00	1728
Lift elements into position		2			864.00	1728
Final positioning and tack welding		2			432.00	864
Final Welding		2			1296.00	2592
PF Blowback N2 Buffer Tank (38.4'x8.8'x8.8)	91	1			240.00	240
Pulverized Feedstock Distribution Hopper Exhaust Collector (22.9'x16.4'x36.1')	62	1			100.00	100
High Pressure Soot Blower						294
Handle Soot Blowers						
High Pressure Soot Blower C, D, E, F, G, H (30'x3'x4')	2	6			10.00	60
High Pressure Soot Blower C, D, E, F, G, H (30'x3'x4')		6			24.00	144
High Pressure Shutoff Valve 6" Diameter Weld				12	7.50	90

7.3.6 Equipment estimate sheet 6

7.3.6 Equipment Estimate Sheet 6 Description	Wt. (ton)	Qty Hdl/Set	Qty Field Jt	Qty Bolt Up	MH/Qty	BM-MH
Gasifier and SGC Pressure Vessel						
Gasifier and SGC Pressure Vessel (GP1) (62'x23'x22')	360					240
Receive, Jack, Transport, Set & Shape Adjustment		1			240.00	240
Gasifier and SGC Pressure Vessel (GP2) (60'x23'x21')	362					4120
Receive, Jack, Transport, Set & Shape Adjustment		1			240.00	240
Weld Prep; scaffold, Joint prep/fit, backing strip, machine set up; GP1/GP2		1	1		1600.00	1600
Weld Outside, Back Gouge & Weld Inside; 16'-9", WT=4.69"			1		1680.00	1680
Craft Support Preheat & PWHT			1		600.00	600
Gasifier and SGC Pressure Vessel (GP3) (49'x22'x23')	316					4120
Receive, Jack, Transport, Set & Shape Adjustment		1			240.00	240
Weld Prep; scaffold, Joint prep/fit, machine set up; GP1/GP2 to GP3		1	1		1600.00	1600
Weld Outside, Back Gouge & Weld Inside; 16'-9", WT=4.45"			1		1680.00	1680
Craft Support Preheat & PWHT			1		600.00	600
Gasifier and SGC Pressure Vessel (GP4) (57'x23'x23')	353					4360
Receive, Jack, Transport, Set & Shape Adjustment		1			240.00	240
Weld Prep; scaffold, Joint prep/fit, backing strip, machine set up; GP1/GP2/GP3 to GP4		1	1		1600.00	1600
Weld Outside, Back Gouge & Weld Inside; 16'-9", WT=4.57"			1		1680.00	1680
Craft Support Preheat & PWHT			1		600.00	600
Manhole Cover 30" Diameter			5		20.00	100
Gas Temperature Mounting Blind Flange 13" Diameter			5		20.00	100
Temperature Mounting Blind Flange 3" Diameter			21		1.90	40
GP1.GP2, GP3, GP4						650
Bottom support ring (Octagon-shaped) (20'x20'x8')	61	1		48	3.33	160
Stopper BW		1		8	10.00	80
Stopper; 7.9' 0.6'' Thk		1	1		10.00	10
Set Vessel		1			400.00	400

7.3.7 Equipment estimate sheet 7

7.3.7 Equipment Estimate Sheet 7	Wt. (ton)	Qty Hdl/Set	Qty Field Jt	Qty Bolt Up	MH/Qty	BM-MH
Description						
Gasifier and SGC internals						
Gasifier and SGC Water Wall, Economizer, 1ry-SH (GW1); Membrane Wall 1.5"						1924
Receive, Transport & Install		1			160.00	160
Water Wall; SA-213T11+SA-213T310; 1.5" Dia x 0.181" WT BW Tig			193		3.52	679
Water Wall-MH Panel; SA-213T11+SA-213T310, 1.5" Dia x 0.181" WT BW Tig			32		3.52	113
Water Wall-SB Panel; SA-213T11+SA-213T310; 1.5" Dia x 0.181" WT BW Tig			14		3.52	49
Economizer Element Tube; SA-213T11+SA-213TP10; 1.772" Dia x 0.181" WT BW			104		3.68	383
Economizer Hanger Tube; SA-213T11+SA-213TP10; 1.913" Dia x 0.339" WT BW;			26		3.68	96
Fit & Weld Filler Bar (Fixing Piece) @ Tube Welds GW1/GW2			187		1.60	299
Fit & Weld Filler Bar (Fixing Piece) @ Tube Welds; FW GW1/GW2			8		1.92	15
Fit & Weld Filler Bar (Fixing Piece) @ Tube Welds; FW GW1/GW2 MH Panel			30		1.60	48
Fit & Weld Filler Bar (Fixing Piece) @ Tube Welds; FW GW1/GW2 MH Panel			4		1.92	8
Fit & Weld Filler (Panel End bar); BW GW1/GW2			77		0.96	74
Gasifier and SGC Water Wall, 1ry-SH, 2ry-SH, Evap (GW2); Membrane Wall 1.5"						2102
Receive, Transport & Install		1			160.00	160
Water Wall; SA-213T11+SA-213T310; 1.5" Dia x 0.181" WT BW; Tig			209		3.52	736
Water Wall-MH Panel; SA-213T11+SA-213T310; 1.5" Dia x 0.181" WT B Tig			32		3.52	113
Water Wall-SB Panel; SA-213T11+SA-213T310; 1.5" Dia x 0.181" WT BW Tig			14		3.52	49
1ry SH Element Tube; SA-213TO347HFG; 1.772" Dia x 0.150" WT BW; Tig			104		4.60	478
Economizer Hanger Tube; SA-213T11+SA-213TP10; 1.913" Dia x 0.339" WT BW;			26		3.68	96
Fit & Weld Filler Bar (Fixing Piece) @ Tube Welds; FW GW2/GW3			203		1.60	325
Fit & Weld Filler Bar (Fixing Piece) @ Tube Welds GW2/GW3			8		1.92	15
Fit & Weld Filler Bar (Fixing Piece) @ Tube Welds' GW2/GW3 MH Panel			30		1.60	48
Fit & Weld Filler Bar (Fixing Piece) @ Tube Welds; FW GW2/GW3 MH Panel			4		1.92	8
Fit & Weld Filler (Panel End bar) BW GW2/GW3			77		0.96	74

7.3.8 Equipment estimate sheet 8

7.3.8 Equipment Estimate Sheet 8	Wt. (ton)	Qty Hdl/Set	Qty Field Jt	Qty Bolt Up	MH/Qty	BM-MH
Description						
Gasifier and SGC Water Wall, Evaporator (GW3); Membrane Wall 1.5" Dia						1924
Receive, Transport & Install		1			160.00	160
Water Wall; SA-213T11+SA-213T310; 1.5" Dia x 0.181" WT BW; (Heliarc)			232		3.52	817
Evaporator Element Tube; SA-213TO347HFG; 1.772" Dia x 0.150" WT BW; (Heliarc)			104		4.60	478
Economizer Hanger Tube; SA-213T11+SA-213TP10; 1.913" Dia x 0.339" WT BW; Ti			26		3.68	96
Fit & Weld Filler Bar (Fixing Piece) @ Tube Welds; FW GW3/GW4			224		1.60	358
Fit & Weld Filler Bar (Fixing Piece) @ Tube Welds; FW GW3/GW4			8		1.92	15
Gasifier and SGC Water Wall (GW4); Membrane Wall 1.5" Dia						1350
Receive, Transport & Install		1			160.00	160
Water Wall; SA-213T11+SA-213T310; 1.5" Dia x 0.181" WT BW; (Heliarc) GW4/GW5			232		3.52	817
Fit & Weld Filler Bar (Fixing Piece) @ Tube Welds; FW GW4/GW5			224		1.60	358
Fit & Weld Filler Bar (Fixing Piece) @ Tube Welds; FW GW4/GW5			8		1.92	15
Gasifier and SGC Water Wall (GW5); Membrane Wall 1.5" Dia						1350
Receive, Transport & Install		1			160.00	160
Water Wall; SA-213T11+SA-213T310; 1.5" Dia x 0.181" WT BW ;(Heliarc) GW5			232		3.52	817
Fit & Weld Filler Bar (Fixing Piece) @ Tube Welds; FW GW5			224		1.60	358
Fit & Weld Filler Bar (Fixing Piece) @ Tube Welds; FW GW5			8		1.92	15
GW1, GW2, GW3, GW4 & GW5						4070
Gasifier Cooling Tube; SA-213T22; 1.913" Dia x 0.315" WT Boulet Header Nozzle			18		3.68	66
Gasifier Cooling Tube; SA-213T22; 1.913" Dia x 0.315" WT BW; Terminal Tube			18		3.68	66
Gasifier Cooling Tube; SA-213T22; 1.913" Dia x 0.315" WT Abterminal Tube GW2			18		3.68	66
Fit & Weld Filler Bar (Fixing Piece) @ Tube Welds; FW SB Panel, GW2, GW3			12		1.60	19
Fit & Weld Filler Bar (Fixing Piece) @ Tube Welds; FW SB Panel, GW2, GW3			4		1.92	8
Fit & Weld Filler (Panel Endbar) GW2/GW3			55		0.96	52
Fit & Weld Filler Bar (Fixing Piece) @ Tube Welds; FW SB Panel, GW2, GW3			12		1.60	19
Fit & Weld Filler Bar (Fixing Piece) @ Tube Welds; FW SB Panel, GW2, GW3			4		1.92	8
Fit & Weld Filler (Panel Endbar); SA-240-309S; 0.236" thk x 54.5" L BW			55		0.96	52
Fit & Weld Water Wall Vertical Backstay; SA-387-22-1 BW			1718		1.50	2578
Fit & Weld Keeper Lug, SA-387-22-1 0.5" thk x 11.8" L BW			283		1.50	425
Fit & Weld Gasifier Water Wall Stopper; SA-387-22-1 BW			144		1.50	216
Fit & Weld Erosion Protector; SA-240-310S 0.2" thk x 3.6" L FW Evaporator			374		0.65	243
Fit & Weld Erosion Protector; SA-240-310S 0.2" thk x 1.0" L FW Economizer Hanger Tube			208		0.65	135
Drain Pipe, Air Vent Pipe, Water Pipe & Nitrogen Pipe; SA-335P22 Fit & Weld BW			10		11.55	116

7.3.9 Equipment estimate sheet 9

7.3.9 Equipment Estimate Sheet 9	Wt. (ton)	Qty Hdl/Set	Qty Field Jt	Qty Bolt Up	MH/Qty	BM-MH
Description						
Gasifier Start-up Burner A, B (29.2'x4.9'x4.3')	**8**	**2**			**60.00**	**120**
Gasification Solids Melting Burner A.B (28.5'x8.5'x6.1')	**7**	**2**			**60.00**	**120**
Char Burner A, B, C, D (21.3'x4.6'x4.6')	**7**	**4**			**60.00**	**240**
Pulverized Feedstock Distributor for Reductor Burner (4'x4'x4')	**1**	**1**			**20.00**	**20**
Pulverized Feedstock Distributor for Combustor Burner (4'x4'x4')	**1**	**1**			**20.00**	**20**
Mechanical Equipment						**9182**
Gasification Solids Lock Hopper (25'x15'x16')	64	1			160.00	160
Weld Stopper 1.26" The x 55.12" L BW			220		1.35	298
Weld Manhole Nozzle 26" Diameter		1	82		0.49	40
Maintenance Hoist for Feedstock Pulverize (1/3) 60 HP	4	1			180.00	180
Maintenance Hoist for Feedstock Pulverize (2/3) 60 HP	4	3			180.00	540
Maintenance Hoist for Feedstock Pulverize (3/3) 60 HP	5	2			180.00	360
Gasification Solids Cooling Water Cooler (10.1'x5.1'x10.7')	9	1			160.00	160
Feedstock Pulverize A (17.4'x17.6'x28.1')	315					
Separator Top Assembly	34	1	1		320.00	320
Middle Housing Assembly	35	1	1		320.00	320
Lower Housing Assembly	27	1	1		200.00	200
Grinding Table Assembly	41	1	1		280.00	280
Roller Journal Assembly	28	3	3		280.00	840
Roller Cover Assembly	13	3	3		160.00	480
Planetary Gear Mounting Plate	10	1	1		120.00	120
Spillage Chute Assembly	0	1	1		40.00	40
Upper Housing Operation Ladder	0	23	1		0.52	12
Coal Chute	1	1	2		40.00	80
Planetary Reduction Gera	39	1	1		320.00	320
Coal Chute	1	1	2		40.00	80
Coal Chute	1	1	2		40.00	80
Coal Chute	1	1	2		30.00	60
Coal Chute	1	1	2		30.00	60
Feedstock Pulverize -A Motor	10	1			220.00	220

7.3.10 Equipment estimate sheet 10

7.3.10 Equipment Estimate Sheet 10	Wt. (ton)	Qty Hdl/Set	Qty Field Jt	Qty Bolt Up	MH/Qty	BM-MH
Description						
Feedstock Pulverize B (17.4'x17.6'x28.1')	315	1	1		320	320
Separator Top Assembly (17.4'x17.4'x10')	34	1	1		320	320
Middle Housing Assembly (18.1'x18.1'x12.8')	35	1	1		200	200
Lower Housing Assembly (15.8'x17.6'x5,3')	27	1	1		280	280
Grinding Table Assembly (13.0'x13.0'x5.7')	41	3	3		280	840
Roller Journal Assembly (10.0'x10.0'x7.4')	28	3	3		160	480
Roller Cover Assembly (9.2'x6.6'x16.1')	13	1	1		120	120
Planetary Gear Mounting Plate (11.2'x8.3'x2.0')	10	1	1		40	40
Spillage Chute Assembly (2.8'x1.4'x5')	0.5	23	1		0.52	12
Upper Housing Operation Ladder (3'x3'x23')	0.2	1	2		40	80
Coal Chute (3.6'x3.6'x12.5')	1	1	1		320	320
Planetary Reduction Gera (8.2'x10.1'x7')	39	1	2		40	80
Coal Chute (3.8'x3.8'x10.5')	1	1	2		40	80
Coal Chute (3.5'x3.5'x12')	1	1	2		30	60
Coal Chute (3.6'x3.6'x4')	1	1	2		30	60
Coal Chute (4.9'x4.9'x5.5')	1	1			220	220
Feedstock Pulverize -B Motor	10	1			20	20
Auxiliary Steam Drain Recovery Pump Centrifugal 7.4 HP (3.8'x1.4'x1.8')	0.1	1	1		60	60
Gasifier Flash Pipe Silencer (7'x7'x17')	3	1	1		100	100
Gasifier Common Silencer (18'x18'x16')	20	1	1		120	120
Maintenance Hoist for Gasifier 30 HP (16.8'x5.0'x7.7')	7	1	1		120	120
Feedstock Drying Gas Stack (16.8' Diameterx8.4'x8.3')	86	1	1		1656.0	1656
Gasification Solids Discharge Conveyor (74.3'x11.4'x26.2')	118	1			1280.0	1280
Gasifier Circulation Pump Cooler A, B & Water Cooler	3	3			23.33	70
Feedstock Conveyer Room Ventilation Fan A, B		2			30.00	60
Feedstock Pulverize & Pulverized Feedstock Feed System		1			6170.00	6170

7.3.11   Equipment estimate sheet 11

7.3.11 Equipment Estimate Sheet 11	Wt. (ton)	Qty Hdl/Set	Qty Field Jt	Qty Bolt Up	MH/Qty	BM-MH
Description						
Pulverized Feedstock Drying System						3950
Pulverized Feedstock Drying Blower (16.6'x18'x21.6')	25.4				800	800
Pulverized Feedstock Drying Blower Inlet Air Filter (12'x12'x1')	0.3	1			30	30
Pulverized Feedstock Drying Blower Suction Silencer (12'x19'x12')	12.1	1			380	380
Pulverized Feedstock Drying Blower Discharge Silencer (12'x19'x12')	12.1	1			380	380
Powdered Activated Carbon Silo/Feeder (14'Dx14'x70')	45.0	1	2	1	1320	1320
Pulverize Inlet Drying Cold Gas Flow Control Damper A	1.4	1			60	60
Pulverize Inlet Drying Cold Gas Flow Control Damper B	1.4	1			60	60
Pulverize Inlet Drying Hot Gas Flow Control Damper A	2.4	1			80	80
Pulverize Inlet Drying Hot Gas Flow Control Damper B	2.4	1			80	80
Pulverized Feedstock Drying Blower Inlet Louver Damper	2.6	1			80	80
Pulverized Feedstock Drying Blower Inlet Shut Off Damper	4.6	1			140	140
Pulverized Feedstock Drying Blower Air Inlet Shut Off Damper	4.6	1			140	140
Pulverize Inlet Drying Cold Gas Flow Shut Off Damper A	2.1	1			80	80
Pulverize Inlet Drying Cold Gas Flow Shut Off Damper B	2.1	1			80	80
Pulverize Inlet Drying Hot Gas Flow Shut Off Damper A	3.7	1			120	120
Pulverize Inlet Drying Hot Gas Flow Shut Off Damper B	3.7	1			120	120

7.3.12   Equipment estimate sheet 12

7.3.12 Equipment Estimate Sheet 12	Wt. (ton)	Qty Hdl/Set	Qty Field Jt	Qty Bolt Up	MH/Qty	BM-MH
Description						
Pulverized Feedstock Burner						1520
Pulverized Feedstock Reductor Burner A	4.0	1			140	140
Pulverized Feedstock Reductor Burner B	4.0	1			140	140
Pulverized Feedstock Reductor Burner C	4.0	1			140	140
Pulverized Feedstock Reductor Burner D	4.0	1			140	140
Pulverized Feedstock Combustor Burner A	6.7	1			240	240
Pulverized Feedstock Combustor Burner B	6.7	1			240	240
Pulverized Feedstock Combustor Burner C	6.7	1			240	240
Pulverized Feedstock Combustor Burner D	6.7	1			240	240
Char Recovery and Feed System						1720
Char Distribution Hopper Fluidization Chamber A						
Char Distribution Hopper Fluidization Chamber B						
Char Distribution Hopper Fluidization Chamber C						
Char Distribution Hopper Fluidization Filter A	0.2	1			20	20
Char Distribution Hopper Fluidization Filter B	0.2	1			20	20
Char Distribution Hopper Fluidization Filter C	0.2	1			20	20
Char Distribution Hopper Fluidization Filter Retainer A	2.0	1	148		0.54	80
Char Distribution Hopper Fluidization Filter Retainer B	2.0	1	148		0.54	80
Char Distribution Hopper Fluidization Filter Retainer C	2.0	1	148		0.54	80
Char Pipe Converger	0.7	1			40	40
Char Distributor for Char Burner	0.8	1			40	40
Irregular Char Removal Hopper	13.5	1			520	520
Irregular Char Removal Hopper Element	0.1					
Irregular Char Collection Tank	4.2	1			140	140
Char Level Meter Detector Cooling Fan	4.2	1			140	140
Char Level Meter Detector Cooling Fan Motor	0.6					
Char Level Meter Detector Cooling Fan Inlet Damper	0.3					
Char Level Meter Detector Cooling Fan Silencer	0.1					
Char Level Meter Detector Cooling Fan Protect Metallic Mesh	0.0					
Char Distribution Hopper-A Inlet Gastight Valve	4.2	1	2	2	70.00	140
Char Distribution Hopper-B Inlet Gastight Valve	4.2	1	2	2	70.00	140
Char Distribution Hopper-C Inlet Gastight Valve	4.2	1	2	2	70.00	140
CHAR FLOW CONTROL VALVE	2.0	1	2	2	60.00	120

7.3.13 Equipment estimate sheet 13

7.3.13 Equipment Estimate Sheet 13				Qty	Qty	Qty		
Description			Wt. (ton)	Hdl/Set	Field Jt	Bolt Up	MH/Qty	BM-MH
Gasification Solids Disposal System								5980
Gasification Solids Make-up Water Pump	Centrifgual	100 HP	1.7	1			80	80
Gasification Solids Cooling Water Head Tank			1.1	1			40	40
Chemical Injection Tank			0.3	1			20	20
Gasification Solids Cooling Water Circulation Pump	Centrifgual	250 HP	3.1	1			200	200
Gasification Solids Water Circulation Pump A	Centrifugal	300 HP	11.0	3			300	900
Gasification Solids Water Circulation Pump B	Centrifugal	300 HP	11.0	3			300	900
Gasification Solids Water Cooler A			97.0	1			640	640
Gasification Solids Water Cooler B			97.0	1			640	640
Gasification Solids Relay Conveyor			209.3					
Gasification Solids Distribution Damper			22.7					
Gasification Solids Lock Hopper Feed Water Tank			15.4	1			40	40
Gasification Solids Lock Hopper Feedwater Pump A	Centrifugal	375 HP	11.3	3			360	1080
Gasification Solids Lock Hopper Feedwater Pump B	Centrifugal	375 HP	11.3	3			360	1080
Gasification Solids Disposal System Drain Pump A	Slurry	20 HP	0.6	3			60	180
Gasification Solids Disposal System Drain Pump B	Slurry	20 HP	0.6	3			60	180

Chapter 8

Refinery equipment and storage tank labor estimates

8.1 Introduction

This chapter provides labor estimates for:

- petroleum refinery equipment
- API 650 steel welded storage tank

The labor estimate provides the basis for the project schedule. The estimator uses the labor estimate to determine crew craft and makeup requirements, project duration, and to develop Level I and Level II schedules for the project.

8.2 Refinery equipment estimate

8.2.1 Refinery equipment installation man-hours activity

	BM-MH	PF-MH	Total MH
Total man-hours	18,840	2100	20,940
Vessels/columns	5584		5584
Tanks	280		280
Reactors	2800		2800
Shell and tube heat exchanger	1380		1380
Pumps	2068		2068
Recycle compressor no. 1, reciprocating 1500 HP	1980	800	2780
Recycle compressor no. 2, reciprocating 1500 HP	1980	800	2780
Rich solvent hydraulic turbine, 1292 HP	2460	500	2960
Miscellaneous equipment and special specialty items	308		308

Industrial Construction Estimating Manual. DOI: https://doi.org/10.1016/B978-0-12-823362-7.00008-9

8.3 Refinery equipment bid breakdown

	BM-MH	PF-MH
Vessels/columns	**5584**	
Solvent absorber 8'-6" D × 69'-0"	1620	
Rich solvent flash drum 7'-0" D × 24'-0"	376	
Solvent regenerator 8'-0" D × 72'-6"	1032	
Solvent regenerator reflux drum 5'-0" D × 15'-0"	220	
Solvent regenerator reboiler steam condensate drum 2'-6" D × 8'-0"	60	
Water makeup drum 1'-0" D × 2'-8"	32	
Solvent sump drum 4'-0" D × 12'-0"	176	
Recycle compressor suction drum 6'-6" D × 12'-0"	472	
Reactor steam drum 11'-0" D × 33'-0"	664	
Mixed alcohol absorber 5'-6" D × 24'-0"	884	
Absorber water feed drum 2'-6" D × 7'-6"	48	
Tanks	**280**	
Solvent-storage tank 12'-0" D × 24"-0"	280	
Reactors	**2800**	
Ethanol synthesis reactor (12'-6" D × 25'-9")	700	
Ethanol synthesis reactor (12'-6" D × 25'-9")	700	
Ethanol synthesis reactor (12'-6" D × 25'-9")	700	
Ethanol synthesis reactor (12'-6" D × 25'-9")	700	
Shell and tube heat exchanger	**1380**	
Lean solvent cooler	80	
Solvent regenerator reboiler	40	
Ethanol synthesis feed/effluent exchanger	320	
Reactor feed heater	60	
Recycle compressor spillback cooler	100	
Absorber feed trim cooler	240	
Lean/rich solvent exchangers	240	
Solvent regenerator OVHD condenser	60	
Absorber feed cooler	240	
Pumps	**2068**	
Lean solvent booster pumps 150 HP	480	
Lean solvent pumps 1750 HP	680	
Solvent regenerator reflux pumps 3 HP	40	
Absorber water wash pumps 3 HP	40	
Solvent makeup pump 3 HP	20	
Water makeup pump 7.5 HP	48	
Solvent sump drum pump 3 HP	20	
Steam drum circulation pump 75HP	540	
Sulfating agent injection pump 3HP	40	
Absorber water feed pump 25HP	160	
Recycle compressor no. 1, reciprocating 1500 HP	**1980**	**800**
Recycle compressor no. 2, reciprocating 1500 HP	**1980**	**800**
Rich solvent hydraulic turbine, 1292 HP	**2460**	**500**
Miscellaneous equipment and special specialty items	**308**	
Solvent makeup filet, cartridge	48	
Solvent filter package, skid 16'L × 10'W	140	
Antifoam injection package, chemical skid	120	

8.3.1 Vessels/columns sheet 1

Scope	Wt. (ton)	Qty Hdl/ Set	Qty Field Jt	Qty Boltup	MH/ Qty	BM- MH
Solvent absorber 8'-6" D × 69'-0"	**319.7**					**1620**
Unload, handle, haul up to 2000', rig, set, and align, make up Fdn AB	319.7	1			1.88	600
Install platforms and ladders		1			200	200
Remove and replace manway cover (24" 300 removable-davit)			1	1	40	40
Install double downflow valve trays (16 trays)		16			40	640
Install demisting pads (single grid—support, pad, grid-top)		1			60	60
Vortex breaker		1			32	32
Packing (pall rings)		1			48	48
Rich solvent flash drum 7'-0" D × 24'-0"	**14.9**					**376**
Unload, handle, haul up to 2000', rig, set, and align, make up Fdn AB	14.9				10.7	160
Install platforms and ladders		1			48	48
Remove and replace manway cover (24" 300 Hinged)		1			24	24
Inlet box		1			60	60
Vortex breaker		1			24	24
Install demisting pads (single grid—support, pad, grid-top)		1			60	60
Solvent regenerator 8'-0" D × 72'-6"	**39**					**1032**
Unload, handle, haul up to 2000', rig, set, and align, make up Fdn AB	39	1			8.2	320
Install platforms and ladders		1			160	160
Remove and replace manway cover (24" 300 removable-davit)			1	1	40	40
Install double downflow valve trays (12 trays)		12			32	384
Install demisting pads (single grid—support, pad, grid-bottom)		1			48	48
Vortex breaker		1			32	32
Packing (pall rings)		1			48	48
Solvent regenerator reflux drum 5'-0" D × 15'-0"	**2.6**					**220**
Unload, handle, haul up to 2000', rig, set, and align, make up Fdn AB	2.6	1			30.8	80
Install platforms and ladders		1			60	60
Remove and replace manway cover (24" 300 Hinged)		1			24	24

(Continued)

(Continued)

Scope	Wt. (ton)	Qty Hdl/ Set	Qty Field Jt	Qty Boltup	MH/ Qty	BM- MH
Install demisting pads (single grid–support, pad, grid-top)		1			32	32
Vortex breaker		1			24	24
Solvent regenerator reboiler steam condensate drum 2'-6" D × 8'-0"	1.5					**60**
Unload, handle, haul up to 2000', rig, set, and align, make up Fdn AB	1.5	1			60	60
Water makeup drum 1'-0" D × 2'-8"	**0.2**					**32**
Unload, handle, haul up to 2000', rig, set, and align, make up Fdn AB	0.2	1			32	32
Solvent sump drum 4'-0" D × 12'-0"	**2.2**					**176**
Unload, handle, haul up to 2000', rig, set, and align, make up Fdn AB	2.2	1			36.4	80
Install platforms and ladders		1			40	40
Remove and replace manway cover (24" 300 Hinged)		1			24	24
Vortex breaker		1			16	16

8.3.2 Vessels/columns/tank sheet 2

Scope	Wt. (ton)	Qty Hdl/ Set	Qty Field Jt	Qty Boltup	MH/ Qty	BM- MH
Recycle compressor suction drum 6'-6" D × 12'-0"	**45.4**					**472**
Unload, handle, haul up to 2000', rig, set, and align, make up Fdn AB	45.4	1			7.93	360
Install platforms and ladders		1			48	48
Remove and replace manway cover (24" 300 Hinged)		1			24	24
Install demisting pads (single grid–support, pad, grid-top)		1			40	40
Reactor steam drum 11'-0" D × 33'-0"	**293.3**	1				**664**
Unload, handle, haul up to 2000', rig, set, and align, make up Fdn AB	293.3	1			1.8	520
Install platforms and ladders		1			120	120
Remove and replace manway cover (24" 300 Hinged)		1			24	24
Mixed alcohol absorber 5'-6" D × 24'-0"	**53.3**	1				**884**

(Continued)

(Continued)

Scope	Wt. (ton)	Qty Hdl/ Set	Qty Field Jt	Qty Boltup	MH/ Qty	BM- MH
Unload, handle, haul up to 2000', rig, set, and align, make up Fdn AB	53.3	1			7.13	380
Install platforms and ladders		1			120	120
Remove and replace manway cover (24" 300 removable-davit)		1			40	40
Install double downflow valve trays (12 trays)		12			24	288
Install demisting pads (single grid—support, pad, grid-top)		1			32	32
Packing (pall rings)		1			24	24
Absorber water feed drum 2'-6" D × 7'-6"	**1,1**					**48**
Unload, handle, haul up to 2000', rig, set, and align, make up Fdn AB	1.1	1			36.4	40
Vortex breaker		1			8	8
Tanks						
Solvent-storage tank 12'-0" D × 24"-0"	**13.1**					**280**
Unload, handle, haul up to 2000', rig, set, and align, make up Fdn AB	13.1	1			21.4	280

8.3.3   Reactors sheet 3

Scope	Wt. (ton)	Qty Hdl/ Set	Qty Field Jt	Qty Boltup	MH/ Qty	BM- MH
Ethanol synthesis reactor (12'-6" D × 25'-9"	**309**					**700**
Unload, handle, haul up to 2000', rig, set, and align, make up Fdn AB	309	1			1.94	600
Install platforms and ladders		1			100	100
Ethanol synthesis reactor (12'-6" D × 25'-9")	**309**					**700**
Unload, handle, haul up to 2000', rig, set, and align, make up Fdn AB	309	1			1.94	600
Install platforms and ladders		1			100	100
Ethanol synthesis reactor (12'-6" D × 25'-9")	**309**					**700**
Unload, handle, haul up to 2000', rig, set, and align, make up Fdn AB	309	1			1.94	600
Install platforms and ladders		1			100	100

(*Continued*)

(Continued)

Scope	Wt. (ton)	Qty Hdl/ Set	Qty Field Jt	Qty Boltup	MH/ Qty	BM- MH
Ethanol synthesis reactor (12'-6" D × 25'-9")	309					700
Unload, handle, haul up to 2000', rig, set, and align, make up Fdn AB	309	1			1.94	600
Install platforms and ladders		1			100	100

8.3.4 Shell and tube heat exchanger sheet 4

Scope	Wt. (ton)	Qty Hdl/ Set	Qty Field Jt	Qty Boltup	MH/ Qty	BM- MH
Lean solvent cooler	18					80
Unload, handle, haul up to 2000', rig, set, and align, make up Fdn AB	18	1			4.44	80
Solvent regenerator reboiler	5					40
Unload, handle, haul up to 2000', rig, set, and align, make up Fdn AB	5	1			8	40
Ethanol synthesis feed/effluent exchanger	100					320
Unload, handle, haul up to 2000', rig, set, and align, make up Fdn AB	100	1			3.2	320
Reactor feed heater	6.5					60
Unload, handle, haul up to 2000', rig, set, and align, make up Fdn AB	6.5	1			9.23	60
Recycle compressor spillback cooler	38					100
Unload, handle, haul up to 2000', rig, set, and align, make up Fdn AB	38	1			2.63	100
Absorber feed trim cooler	90					240
Unload, handle, haul up to 2000', rig, set, and align, make up Fdn AB	90	1			2.67	240
Lean/rich solvent exchangers	90					240
Unload, handle, haul up to 2000', rig, set, and align, make up Fdn AB	90	1			2.67	240
Solvent regenerator OVHD condenser						60
Unload, handle, haul up to 2000', rig, set, and align, make up Fdn AB		1			60	60
Absorber feed cooler						240
Unload, handle, haul up to 2000', rig, set, and align, make up Fdn AB		1			240	240

8.3.5   Pumps sheet 5

Scope	Wt. (ton)	Qty Hdl/ Set	Qty Field Jt	Qty Boltup	MH/ Qty	BM- MH
Lean solvent booster pumps 150 HP						**480**
Unload, handle, haul up to 2000', rig, set, and align, couple and grout		2			240	480
Lean solvent pumps 1750 HP						**680**
Unload, handle, haul up to 2000', rig, set, and align, couple and grout		2			340	680
Solvent regenerator reflux pumps 3 HP						**40**
Unload, handle, haul up to 2000', rig, set, and align, couple and grout		2			20	40
Absorber Water wash pumps 3 HP						**40**
Unload, handle, haul up to 2000', rig, set, and align, couple and grout		2			20	40
Solvent makeup pump 3 HP						**20**
Unload, handle, haul up to 2000', rig, set, and align, couple and grout		1			20	20
Water makeup pump 7.5 HP						**48**
Unload, handle, haul up to 2000', rig, set, and align, couple and grout		2			24	48
Solvent sump drum pump 3 HP						**20**
Unload, handle, haul up to 2000', rig, set, and align, couple and grout		1			20	20
Steam drum circulation pump 75HP						**540**
Unload, handle, haul up to 2000', rig, set, and align, couple, and grout		3			180	540
Sulfating agent injection pump 3 HP						**40**
Unload, handle, haul up to 2000', rig, set, and align, couple and grout		2			20	40
Absorber water feed pump 25HP						**160**
Unload, handle, haul up to 2000', rig, set, and align, couple and grout		2			80	160

8.3.6 Recycle compressor no. 1 sheet 6

Scope	Wt. (ton)	Qty Hdl/ Set	Qty Field Jt	Qty Boltup	MH/ Qty	BM- MH
Recycle compressor no. 1, reciprocating 1500 HP						1980
Frame and gear	1				340	340
34'-5" Cylinder	1				180	180
22'-0" Cylinder	1				120	120
First-stage suction damper	1				60	60
First-stage discharge damper	1				60	60
Extended bearing and postal	1				60	60
Lube oil consul 2 ea.	1				120	120
Moister separator 3 ea.	1				180	180
Full maintenance deck	1				240	240
IC pipe—interstate piping spools between pulsation bottles and coolers	1				800	
Rotor and extension shaft	1				240	240
Motor stator	1				180	180

8.3.7 Recycle compressor no. 2 sheet 7

Scope	Wt. (ton)	Qty Hdl/ Set	Qty Field Jt	Qty Boltup	MH/ Qty	BM- MH	PF- MH
Recycle compressor No. 2, reciprocating 1500 HP						1980	800
Frame and gear	1				340	340	
34'-5" Cylinder	1				180	180	
22'-0" Cylinder	1				120	120	
First-stage suction damper	1				60	60	
First-stage discharge damper	1				60	60	
Extended bearing and pestal	1				60	60	
Lube oil consul 2 ea.	1				120	120	
Moister separator 3 ea.	1				180	180	
Full maintenance deck	1				240	240	
IC pipe—interstage piping spools between pulsation bottles and coolers	1				800		800
Rotor and extension shaft	1				240	240	
Motor stator	1				180	180	
Motor enclosure (soleplates, inc)	1				200	200	

8.3.8   Rich solvent hydraulic turbine, 1292 HP Sheet 8

Scope	Wt. (ton)	Qty Hdl/ Set	Qty Field Jt	Qty Boltup	MH/ Qty	BM- MH	PF- MH
Rich solvent hydraulic turbine, 1292 HP						2460	500
Turbine/gearbox skid	1				460	460	
Rotor	1				320	320	
High/low-speed cplg spacers and disk packs	1				60	60	
Steam inlet trip and throttle valve	1				80		80
Lube oil skid	1				140	140	
Generator w/sole plates	1				800	800	
Generator panels 2 ea.	1				80	80	
Gear terminal boxes	1				600	600	
IC pipe	1				420		420

8.3.9   Miscellaneous equipment and special specialty items sheet 9

Scope	Wt. (ton)	Qty Hdl/ Set	Qty Field Jt	Qty Boltup	MH/ Qty	BM- MH	PF- MH
Miscellaneous equipment and special specialty items						308	
Solvent makeup filter, cartridge						48	
Unload, handle, haul up to 2000′, rig, set, and align		1			48	48	
Solvent filter package, skid 16′L × 10′W						140	
Unload, handle, haul up to 2000′, rig, set, and align, make up Fdn AB		1			140	140	
Antifoam injection package, chemical skid						120	
Unload, handle, haul up to 2000′, rig, set, and align, make up Fdn AB		1			120	120	

8.4   API 650 oil storage tanks

Estimate for API Std welded steel storage tanks used for petroleum products and other liquid products stored in the petroleum industry.

8.4.1 Tank data

Tank dimensions: 80′ D × 24′ H
 Standard Sheet: 10′L × 8′W
 Tank shell plate: Number of sheets per ring 26
 Ring 1 26 sheets × 0.75″ Thk.
 Ring 2 26 sheets × 0.5625″ Thk.
 Ring 3 26 sheets × 0.4375″ Thk.
 Roof and floor sheets
 Number of roof sheets 63 × 0.5″ Thk.
 Number of floor sheets 63 × 0.5″ Thk.
 Gallons: 902,430 gallons
 Barrels (42 Gallons/BBL) 21,486 BBL

8.5 Tank construction estimate

8.5.1 Introduction

This section provides a general method to erect a welded steel storage tank. Tank erectors each has a particular method, which has been developed by experience that is suitable for their field crews to work economically and provide quality work. To erect tanks that are of sound quality, appearance, and without distortion, the erector must adherer to corrected welding sequences and maintain adequate supervision.

8.5.2 Tank erection bid breakdown

	BM-MH
Activity	
Erect knuckle ring steel welded storage tank	4005
Receive and offload bottom shell and roof plates, and all material	224
Set bottom plates and weld	676
Shell plate erection	1356
Roof erection	1605
Tank testing	144

8.5.3 Bottom plate placement

When tank foundation is complete, the bottom plate will be placed on the foundation and welded in sequence. Plates are placed and welded in the correct sequence to avoid weld distortion.

8.5.4 Bottom plate placement and welding sequence

	Qty	Qty	Qty		
Scope	Hdl/ Set	Field Jt	Boltup	MH/ Qty	BM-MH
Set bottom plates and weld					**676**
Place plates and tack weld	63			2	95
Set and weld center sump	1	1		4	4
Weld rectangular plates, end seams first, lap jt fillet weld	22	792		0	198
Weld outer radial seams of annular plates	12	120		1	60
First ring set and welded; weld first ring to annular plate fillet weld		502.7		0	176
Weld remaining seams of annular plates	12	120		1	60
Weld rectangular and sketch plates together and to annular		112		1	84

8.5.5 Shell plate erection

Shell plates are rigged and set in place using the crane; fit up, tacked, and the vertical seams are welded in place. The shell rings are done ring-by-ring to the top curb angle. Each ring must be completely weld before another ring can be added.

8.5.6 Shell plate erection

	Qty	Qty	Qty		
Scope	Hdl/ Set	Field Jt	Boltup	MH/ Qty	BM-MH
8.5.6 Shell plate erection					**1356**
First ring; rig and set sheets in place	26			2	52
Fit and weld vertical seams (0.75″ BW)	25	200		1	285
Weld scaffold brackets and erect scaffold	1			32	32
Second ring: rig and set sheets, fit, and tack	26			3	65
Set weld buggy and fit and weld vertical seams (0.5625″ BW)	25	200		1	195
Fit and hand weld horizontal seam (0.5625″ BW); second ring to first ring		251		1	207
Weld scaffold brackets and move scaffold to next ring	1			40	40
Third ring: rig and set sheets, fit, and tack	26			3	78
Move weld buggy to next ring and fit and weld vertical seams (0.4375″ BW)	25	200		1	195
Fit and hand weld horizontal seam (0.4375″ BW); third ring to second ring		251		1	207

8.5.7 Roof erection

Scope	Qty Hdl/ Set	Qty Field Jt	Qty Boltup	MH/ Qty	BM- MH
Roof erection					**1605**
Set and weld center post/column	1			16	16
Rig, set, and fit tack knuckle ring to third ring	26			3.1	81
Weld scaffold brackets and move scaffold to next ring	2			40	80
Rig, place fit, and weld wind girder	26	251		0.75	188
Weld vertical seams knuckle ring (0.4375" BW)	25	150		0.975	146
Weld knuckle ring horizontal weld to third ring (0.4375" BW)		251		0.825	207
Rig and place rafters	25			4	100
Attached rafters to center post and knuckle ring			50	1.5	75
Rig, place, and fit roof sheets	63			2	126
Set and weld roof vents	1			24	24
Weld roof sheets	63			6	378
Stairway, handrails, and platforms	1			120	120
Drain pipe	1			40	40
Manhole	1			24	24

8.5.8 Tank testing

					144
Bottom plate welds are tested to ensure bottom has no leaks. Test using a vacuum box	1			64	64
Hydrotest tank; test tank for leakage and foundation for capability of taking tank load	1			80	80

Chapter 9

Circulating fluidized bed combustion (FBC) labor estimates

9.1 Introduction

This chapter provides the reader a basic understanding of the fundamentals and operating relationship between the plant equipment in the combustion process and covers the craft labor for the assembly and field erection required to put the equipment into operation in a circulating fluidized bed boiler. Fluidized bed combustion (FBC) is a combustion technology used to burn solid fuels. Fuel particles are suspended in a hot, bubbling fluidity bed of ash, and other materials, such as sand and limestone, through which jets of air are blown to provide oxygen required for combustion or gasification. The resultant fast and intimate mixing of gas and solids promotes raped heat transfer and chemical reactions within the bed.

The plants are attractive because of the following reasons:

- FBC plants burn low-grade solid fuels, including woody biomass, without expensive fuel preparation.
- FBCs are smaller than equivalent conventuals and have significant advantages over the latter in terms of cost and flexibility.
- FBCs have SO_2 and NO_2 emissions below Federal Standards.

9.2 Combustor bid breakdown

	BM	Laborer	Total MH
Combustor	**8902**	**560**	**9462**
Combustor support structure	40	560	600
Platforms (two levels)	440		440
Cones (9′ × 10′)	100		100
Manifold assembly	1000		1000
Plenum box [fabr in (1) section] (12′ × 4′)	474		474
Panels	1840		1840

(*Continued*)

Industrial Construction Estimating Manual. DOI: https://doi.org/10.1016/B978-0-12-823362-7.00009-0

(Continued)

	BM	Laborer	Total MH
Wall tubes	2700		2700
Roof panels	280		280
Stoker	360		360
Metering bins	696		696
Burners	972		972

9.3 Combustor

9.3.1 Combustor equipment labor hours sheet 1

Scope	Qty	Unit	MH/ Qty	BM	Laborer	Total MH
Combustor support structure				**520**	**40**	**560**
Erect columns	10	ea.	10	100		100
Column field splice	10	ea.	8	80		80
Erect horizontal beams, bolt connections	16	ea.	8.75	140		140
Structural X-brace, bolt connections	16	ea.	10	160		160
Shim, bolt column base plate to fdn	10	ea.	4	40		40
Grout column base plates	10	ea.	4		40	40
Platforms (two levels)				**440**		**440**
Platform structural steel	660	sf	0.23	150		150
Grating	660	sf	0.21	140		140
Handrail	212	lf	0.28	60		60
Stairs	85	lf	0.94	80		80
Ladders	32	lf	0.31	10		10
Cones (9′ × 10′)				**100**		**100**
Field fabr cone sections (2 sections)	2	ea.	30	60		60
Set cones and attach to support structure	2	ea.	20	40		40
Manifold assembly				**1000**		**1000**
Install sand plows (L 1 1/2″)	80	ea.	1.0	80		80
Plug weld nozzle holes (9/32″)	70	ea.	1.1	80		80
Prefabricated air headers	6	ea.	20.0	120		120
Seal weld/butt weld manifolds together	160	lf	1.4	216		216
Install nozzles (1-1/2″)	840	ea.	0.6	504		504
Plenum box [fabr in (1) section] (12′ × 4′)				**474**		**474**
Set plenum and weld to manifold	2	pcs	30	60		60
Seal weld	120	lf	1.2	144		144
Structural weld	25	lf	1.2	30		30

(Continued)

(Continued)

Scope	Qty	Unit	MH/ Qty	BM	Laborer	Total MH
Combustor lower A plenum splices	10	ea.	8	80		80
Combustor middle A plenum splices	10	ea.	8	80		80
Combustor upper A plenum splices	10	ea.	8	80		80

9.3.2   Combustor equipment labor hours sheet 2

Scope	Qty	Unit	MH/ Qty	BM	Laborer	Total MH
Panels				**1840**		**1840**
Lower section (10′ × 13′)						
Set panels (4 ea.), fit up vertical seam, and butt weld	40	ea.	2.5	100		100
Weld lower section to manifolds	46	lf	1.30	60		60
Transition						
Erect transition panels (2 panels/ side)	8	ea.	10.00	80		80
Fit up and weld (4 ea.)	100	lf	1.20	120		120
Fit up and weld (4 ea.)	124	lf	1.13	140		140
Weld transition panels to lower section	46	lf	1.30	60		60
Upper section						
6 panels/ring × 5 rings						
Set panels (6 ea.) × 5	30	ea.	16.00	480		480
Fit up and weld vertical (6 × 5 × 7.2)	216	lf	1.20	260		260
Fit up and weld horizontal (92 × 4)	368	lf	1.14	420		420
Weld upper section to transition	92	lf	1.30	120		120
Wall tubes				**2700**		**2700**
3.5″ × s 80 cs × 50′ × 70 ea. wall tubes rig and place in unit	3500	lf	0.35	1240		1240
Headers embedded in wall—not included in estimate						
Headers embedded in wall—not included in estimate						
Headers embedded in wall—not included in estimate						
3.5″ × s 80 CS field butt weld NDE X-ray	210	ea.	4.67	980		980
3.5″ CS fillet weld exp Jt to lid	70	ea.	2.57	180		180
Spring hanger	10	ea.	6.00	60		60
Hydrotest	1	lot	240	240		240

(Continued)

(Continued)

Scope	Qty	Unit	MH/ Qty	BM	Laborer	Total MH
Roof panels				280		280
Panels						
Set panels (4 ea.)	4	ea.	20	80		80
Fit up and weld panels	166	lf	1.20	200		200

9.3.3 Combustor equipment labor hours sheet 3

Scope	Qty	Unit	MH/ Qty	BM	Laborer	Total MH
Stoker				360		360
LH fixed frame assembly	2	ea.	10	20		20
Center fixed frame assembly	2	ea.	10	20		20
RH fixed frame assembly	2	ea.	10	20		20
Pipe						
4″ straight pipe/pipe Supports	48	lf	1.67	80		80
Fuel distributors 36″ × 10″	4	ea.	5	20		20
Rotating air dampers	4	ea.	10	40		40
Balance dampers	4	ea.	10	40		40
High pressure air headers and nozzles	6	ea.	10	60		60
Manual control damper	6	ea.	10	60		60
Metering bins				696		696
Metering bins	2	ea.	40	80		80
Discharge chute	2	ea.	40	80		80
Metering bin screw	2	ea.	30	60		60
Feed chutes	2	ea.	30	60		60
Expansion joint	2	ea.	50	100		100
Metering bin fuel feed structure	18	Ton	14.2	256		256
Bin inlet slide gate	2	ea.	10	20		20
Inlet metering bin slide gate	2	ea.	10	20		20
Fuel spreaders	2	ea.	10	20		20
Burners				972		972
Overbed burner and wind box	2	ea.	40	80		80
Burner IV .75″ IC piping	60	lf	2	120		120
Underbed burner and wind box	1	ea.	40	40		40
Burner IV .75″ IC piping	30	lf	2	60		60
Fuel gas skid	2	ea.	40	80		80
Underbed burner gas train	2	ea.	40	80		80
Ductwork (24″ dia)	6	pcs	13.3	80		80
Ductwork (18″ dia)	2	pcs	20	40		40
Dampers (18″ dia)	2	ea.	10	20		20
Duct supports	7	ea.	5.7	40		40
Misc. duct, flex, orifice, valve	2	Ls	20	40		40
Seal weld	15	lf	1.3	20		20

(Continued)

(Continued)

Scope	Qty	Unit	MH/Qty	BM	Laborer	Total MH
Buck stays	16	ea.	10	160		160
Attachment clips	280	ea.	0.5	140		140
Sight glass assembly	4	ea.	8	32		32
Doors	2	ea.	10	20		20

9.4 Boiler bid breakdown

	BM	PF	Total MH
Boiler	**5368**	**5192**	**10,560**
Support structure 45 ton	1200		1200
Boiler access steel	1120		1120
Modules	920		920
Boiler outlet/inlet transition	388		388
Expansion joint at inlet/outlet (at combustor and economizer)	160		160
Hoppers	120		120
Drum	60		60
IC piping, BD, vents, drains, and trim at drum		5192	5192
Soot blowers	400		400
Silencers	360		360
Hydrotest	240		240
Stack (3—sections) w/1 platform	400		400

9.5 Boiler

9.5.1 Boiler labor hours sheet 1

Scope	Qty	Unit	MH/Qty	BM	PF	Total MH
Support structure 45 ton				**1200**		**1200**
Erect columns	16	ea.	10	160		160
Column field splice	32	ea.	8	256		256
Erect horizontal beams, bolt connections	64	ea.	5	320		320
Structural X-brace, bolt connections	32	ea.	10	320		320
Shim, bolt column base plate to fen	16	ea.	4.5	72		72
Grout column base plates	16	ea.	4.5	72		72
Boiler access steel				**1120**		**1120**
Grating	4000	sf	0.19	760		760
Handrail	1200	lf	0.25	300		300

(Continued)

(Continued)

Scope	Qty	Unit	MH/ Qty	BM	PF	Total MH
Stairs	61	lf	0.98	60		60
Modules				920		920
SH # 1 box rig and set	1	ea.	80	80		80
SH # 2 box rig and set	1	ea.	80	80		80
Evap #1 box rig and set	1	ea.	80	80		80
Eva #2 box rig and set	1	ea.	80	80		80
Field joint (box to box, box to transition) 5 EA	296	lf	2.03	600		600
Boiler outlet/inlet transition				388		388
Roof, wall, and side panels	8	ea.	20	160		160
Panel clips (3/4" × 10" × 12" lg)	20	ea.	1.20	24		24
Seal weld	90	lf	1.3	120		120
Structural weld	40	lf	1.20	48		48
Bolt and tack weld	20	pcs	1	20		20
Remove lifting brackets (interior)	16	ea.	1	16		16
Expansion joint at inlet/outlet (at combustor and economizer)				160		160
Install expansion joint 8' × 20'	2	ea.	80	160		160
Hoppers				120		120
Set hoppers and bolt to structure	4	ea.	30	120		120
Drum				60		60
Set drum and drum supports	1	ea.	60	60		60

9.5.2 Boiler labor hours sheet 2

Scope	Qty	Unit	MH/ Qty	BM	PF	Total MH
IC piping, BD, vents, drains, and trim at drum					5192	5192
Avg −8" s 80 CS pipe NDE	400	lf	2.2		880	880
BD, vents and drains (2" sch 80 cs)	320	lf	2.5		800	800
Boiler trim at drum (pipe and instruments)	1	lot	340		340	340
Boiler piping (from steam drum to soot blowers) NDE	360	lf	3.2		1152	1152
Superheater welds (10" dia) NDE, PWHT	2	ea.	20		40	40
Boiler piping (downcomers)	900	lf	2.2		1980	1980
Soot blowers				400		400
Install soot blowers	20	ea.	20	400		400
Silencers				360		360
Erect silencer support	2	ea.	60	120		120
Install silencers	2	ea.	20	40		40
Silencer Piping (10" s 40 cs)	60	lf	3.33	200		200

(Continued)

(Continued)

Scope	Qty	Unit	MH/Qty	BM	PF	Total MH
Hydrotest				**240**		**240**
Hydrotest boiler	1	lot	240	240		240
Stack (3—sections) w/1 platform	1	ea.	400	**400**		**400**

9.6 Boiler circulation water bid breakdown

	BM	PF	Total MH
Boiler circulating water		**6048**	**6048**
North/south wall downcomers 10″ s80 900# CS pipe		1083	1083
North/south wall r12″ s80 900# CS pipe		465	465
East wall downcomers 10″ s80 900# CS pipe		1797	1797
East wall risers 12″ s80 900# CS pipe		375	375
West wall downcomers 10″ s80 900# CS pipe		420	420
West wall risers		548	548
Structural support for spring hangers		240	240
Spring hangers		896	896
Shoes		224	224

9.7 Boiler circulation water

Circulating water

9.7.1 Boiler circulating water piping labor hours sheet 1

Scope	Qty	Unit	MH/Qty	BM	PF	Total MH
North/south wall downcomers 10″ s80 900# CS pipe					**1083**	**1083**
Tie in to header-buttwelds, s80 (0.594 WT)	8	ea.	9.5		76	76
Pipe—carbon steel, s80, 3 spools × 10′ L × 2	60	lf	1.2		72	72
Buttwelds, s80 (0.594 WT)	12	ea.	9.5		114	114
Pipe—carbon steel, s80, 1 spools × 20′ L × 2	40	lf	1.2		48	48
Buttwelds, s80 (0.594 WT)	8	ea.	9.5		76	76
Pipe—carbon steel, s80, 3 spools × 14′ L × 2	84	lf	1.2		101	101
Buttwelds, s80 (0.594 WT)	12	ea.	9.5		114	114
Pipe—carbon steel, s80, 3 spools × 14′ L × 2	84	lf	1.2		101	101

(Continued)

(Continued)

Scope	Qty	Unit	MH/ Qty	BM	PF	Total MH
Buttwelds, s80 (0.594 WT)	12	ea.	9.5		114	114
Pipe—carbon steel, s80, 4 spools × 12′ L × 2	96	lf	1.2		115	115
Buttwelds, s80 (0.594 WT)	16	ea.	9.5		152	152
NDE X-ray, s80	68	ea.				
North/south wall risers 12″ s80 900# CS pipe					**465**	**465**
Set headers (loose)	2	ea.	20		40	40
Pipe—carbon steel, s80, 6 spools × 12′ L × 2	144	lf	1.80		259	259
Buttwelds, s80 (.688 WT)	12	ea.	13.80		166	166
NDE X-ray, s80	12	ea.				
East wall downcomers 10″ s80 900# CS pipe					**1797**	**1797**
Tie in to header-buttwelds, s80 (.594 WT)	8	ea.	9.5		76	76
Pipe—carbon steel, s80, 9 spools × 12′ L × 8	864	lf	1.2		1037	1037
Buttwelds, s80 (0.594 WT)	72	ea.	9.5		684	684
NDE X-ray, s80	72	ea.				

9.7.2 Boiler circulating water piping labor hours sheet 2

Scope	Qty	Unit	MH/ Qty	BM	PF	Total MH
East wall risers 12″ s80 900# CS pipe					**375**	**375**
Set headers (loose)	4	ea.	20		80	80
Pipe—carbon steel, s80, 6 spools × 12′ L	72	lf	1.80		130	130
Buttwelds, s80 (0.688 WT)	12	ea.	13.80		166	166
NDE X-ray, s80	12	ea.				
West wall downcomers 10″ s80 900# CS pipe					**420**	**420**
Tie in to header-buttwelds, s80 (0.594 WT)	4	ea.	9.50		38	38
Pipe—carbon steel, s80, 8 spools × 12′ L × 2	192	lf	1.20		230	230
Buttwelds, s80 (0.594 WT)	16	ea.	9.50		152	152
NDE X-ray, s80	16	ea.				
West wall risers					**548**	**548**
Set headers (loose)	4	ea.	20		80	80
Pipe—carbon steel, s80, 6 spools × 14′ L × 2	168	lf	1.80		302	302
Buttwelds, s80 (0.688 WT)	12	ea.	13.80		166	166

(*Continued*)

(Continued)

Scope	Qty	Unit	MH/ Qty	BM	PF	Total MH
NDE X-ray, s80	12	ea.				
Structural support for spring hangers	2	Ton	120		**240**	**240**
Spring hangers	56	ea.	16		**896**	**896**
Shoes	56	ea.	4		**224**	**224**

9.8 Fans bid breakdown

	BM	PF	Total MH
Fans	1160		1160
Combustor OFA FD fan—150 HP	197		197
Combustor UFA FD fan—200 HP	153		153
ID fan (500 HP)	811		811

9.9 Fans

9.9.1 Fans labor hours sheet 1

Scope	Qty	Unit	MH/Qty	BM	PF	Total MH
Combustor OFA FD fan—150 HP				**197**		**197**
Inlet damper (3′ × 9′)	1	ea.	74.6	75		75
Motor	1	ea.	20.0	20		20
Inlet silencer (3′ × 9′ × 9′)	1	ea.	42.0	42		42
Inlet expansion joint (3′ × 9′)	1	ea.	20.0	20		20
Inlet expansion joint (6′ × 4′)	1	ea.	20.0	20		20
Silencer	1	ea.	20.0	20		20
Combustor UFA FD fan—200 HP				**153**		**153**
Inlet damper (3′ × 9′)	1	ea.	42.9	43		43
Motor	1	ea.	20.0	20		20
Inlet silencer (3′ × 9′ × 9′)	1	ea.	30.0	30		30
Inlet expansion joint (3′ × 9′)	1	ea.	20.0	20		20
Inlet expansion joint (6′ × 4′)	1	ea.	20.0	20		20
Silencer	1	ea.	20.0	20		20
ID fan (500 HP)				**811**		**811**
ID fan 8.5 ton	1	ea.	246.5	247		247
Damper	1	ea.	40.0	40		40
Motor	1	ea.	100.0	100		100
Silencer	1	ea.	40.0	40		40
Millwright	1	ea.	384.0	384		384

9.10   Fans—FD, ID, OFA bid breakdown

	BM	PF	Total MH
Fans—FD, ID, OFA ductwork	**1940**		**1940**
FD duct overfire	872		872
FD duct underfired	700		700
ID duct	368		368

9.11   Fans—FD, ID, OFA ductwork

9.11.1   Fans—FD, ID, OFA labor hours sheet 1

Scope	Qty	Unit	MH/Qty	BM	PF	Total MH
FD duct overfire				**872**		**872**
Ductwork (36″ dia)/(18″ dia)	6	pcs	40	240		240
Inlet damper (18″ dia)	3	ea	80	240		240
Seal weld	120	lf	3.27	392		392
FD duct underfire				**700**		**700**
Ductwork	1	ea.	320.0	320		320
Underbed expansion Jt (4′ × 4′)	220	lf	0.5	120		120
Seal weld	1	ea.	260.0	260		260
ID duct				**368**		**368**
ID fan discharge ductwork	1	ea.	120.0	120		120
Expansion jet	1	ea.	120.0	120		120
Seal weld ductwork	80	lf	1.6	128		128

9.12   Economizer/inlet duct/hoppers bid breakdown

	BM	PF	Total MH
Economizer/inlet duct/hoppers	**1520**		**1520**
Support structure	300		300
Economizer access steel	260		260
Economizer	960		960

9.13   SCR/economizer support structure

9.13.1   SCR/economizer support structure labor hours sheet 1

Scope		Unit	MH/Qty	BM	PF	Total MH
Support structure				**300**		**300**
Erect columns	4	ea.	10	40		40
Column field splice	4	ea.	10	40		40

(Continued)

(Continued)

Scope		Unit	MH/ Qty	BM	PF	Total MH
Erect horizontal beams, bolt connections	4	ea.	14	56		56
Structural X-brace, bolt connections	4	ea.	20	80		80
Shim, bolt column base plate to fdn	4	ea.	15	60		60
Grout column base plates	4	ea.	6	24		24
Economizer access steel				**260**		**260**
Access steel	3	ton	14	42		42
Grating	700	sf	0.20	139		139
Handrail	280	lf	0.25	70		70
Ladders	30	lf	0.30	9		9
Economizer				**960**		**960**
Economizer—2 ea. modules	2	ea.	120	240		240
Plenum outlet	1	ea.	80	80		80
4 pcs inlet duct, 208 lf seal weld	208	lf	0.77	160		160
6 ea. soot blowers	6	ea.	60	360		360
4 ea. hoppers	4	ea.	30	120		120

9.14 Multiclonebid breakdown

	BM	PF	Total MH
Multiclone	**1440**		**1440**
Support structure	572		572
Hoppers	748		748
Upper housing	120		120

9.15 Multiclone

9.15.1 Multicone labor hours sheet 1

Scope	Qty	Unit	MH/ Qty	BM	PF	Total MH
Support structure				572		572
Erect columns	6	ea.	10	60		60
Column field splice	6	ea.	9.3	56		56
Erect horizontal beams, bolt connections	4	ea.	14	56		56
Structural X-brace, bolt connections	4	ea.	52	208		208
Shim, bolt column base plate to fen	4	ea.	12	48		48
Grout column base plates	4	ea.	6	24		24
Multiclone access steel	4	ton	14	56		56
Grating	120	sf	0.30	36		36

(Continued)

(Continued)

Scope	Qty	Unit	MH/ Qty	BM	PF	Total MH
Handrail	100	lf	0.25	25		25
Ladders	10	lf	0.30	3		3
Hoppers				**748**		**748**
Set hoppers and weld to housing	2	ea.	54	108		108
Set cones in structure (tubesheets) and weld	8	ea.	80	640		640
Upper housing				**120**		**120**
Erect sides	4	ea.	20	80		80
Fit up and weld vertical (8.4 × 4)	33.6	ea.	1.19	40		40

9.16 Spray dryer bid breakdown

	BM	PF	Total MH
Spray dryer	**3560**		**3560**
Support structure	720		720
4 panels/ring × 11 rings	1440		1440
Stair tower	1100		1100
Penthouse (4 panels)/piping in penthouse (1 lot)	300		300

9.17 Spray dryer

9.17.1 Spray dryer labor hours sheet 1

Scope	Qty	Unit	MH/ Qty	BM	PF	Total MH
Support structure				**720**		**720**
Erect columns	12	ea.	10	120		120
Column field splice	12	ea.	8.3	100		100
Erect horizontal beams, bolt connections	8	ea.	12.5	100		100
Structural X-brace, bolt connections	8	ea.	20	160		160
Shim, bolt column base plate to fdn	12	ea.	13.3	160		160
Grout column base plates	12	ea.	6.7	80		80
4 panels/ring × 11 rings				**1440**		**1440**
Set panels (4 ea.) × 1	44	ea.	10	440		440
Fit up and weld vertical (7′ × 11) × 4	380	elf	1.21	460		460
Fit up and weld horizontal (10 × 20)	240	lf	1.25	300		300
Floor/roof	4	ea.	60	240		240
Stair tower				**1100**		**1100**
Erect columns w/beams	4	ea.	30	120		120
Column field splice	4	ea.	10	40		40

(Continued)

(Continued)

Scope	Qty	Unit	MH/ Qty	BM	PF	Total MH
Platforms w/grating/handrail/ladders	7	ea.	31.4	220		220
Stairs/handrail	180		1.9	340		340
Shim, bolt column base plate to fdn	4	ea.	12	48		48
Grout column base plates	4	ea.	8	32		32
Penthouse (4 panels)/piping in penthouse (1 lot)	1	lot	300	**300**		**300**

9.18 Ductwork—multiclone to spray dryer

9.18.1 Ductwork—multicone to spray dryer labor hours sheet 1

Scope	Qty	Unit	MH/ Qty	BM	PF	Total MH
Ductwork—Multiclone to spray dryer				**1100**		**1100**
Vertical duct	7	ea.	20	140		140
Expansion joint	1	ea.	40	40		40
Weld duct at FJ	6	ea.	36.7	220		220
Horizontal duct	8	ea.	40	320		320
Expansion joint	2	ea.	60	120		120
Weld Duct at FJ	7	ea.	37.1	260		260

9.19 Baghouse bid breakdown

	BM	PF	Total MH
Baghouse	**6160**		**6160**
Support structure	760		760
Hopper	320		320
Baghouse modules manifold, plenum, damper, bags, and conveyor	1980		1980
Conveyors	880		880
Penthouse	2220		2220

9.20 Baghouse

9.20.1 Baghouse labor hours sheet 1

Scope	Qty	Unit	MH/ Qty	BM	PF	Total MH
Support structure				**760**		**760**
Erect columns	12	ea.	10	120		120
Column field splice	12	ea.	10	120		120
Erect horizontal beams, bolt connections	8	ea.	15	120		120
Structural X-brace, bolt connections	8	ea.	20	160		160
Shim, bolt column base plate to fdn	12	ea.	13.3	160		160
Grout column base plates	12	ea.	6.7	80		80
Hopper				**320**		**320**
Set hoppers and bolt to structural	4	ea.	60	240		240
Hopper headers	4	ea.	10	40		40
Top lid/air header	4	ea.	10	40		40
Baghouse modules manifold, plenum, damper, bags, and conveyor				**1980**		**1980**
Set baghouse modules/attach to structural	4	ea.	60	240		240
Install manifold	3	ea.	60	180		180
Fabr and erect inlet/outlet plenum (3 sections × 2), weld to modules	2	ea.	150	300		300
Dampers to each module (8) and (1) bypass damper	36	ea.	18.3	660		660
Install bags in hoppers (300 bags/ hopper)	1200	ea.	0.5	600		600
Conveyors				**880**		**880**
Ash hopper screw conveyor, Austin-MAC	2	ea.	60	120		120
Ash conditioner, ash tech model M-12	2	ea.	90	180		180
Tail and trough	2	ea.	290	580		580
Penthouse				**2220**		**2220**
Penthouse floor sections	10	ea.	10	100		100
Stacking legs/brace assemblies/hold downs	1	lot	120	120		120
Frames	2	ea.	10	20		20
Purlins	375	lf	0.53	200		200
Girts	400	lf	0.5	200		200
Sag rods	8	ea.	10	80		80
Louvers	4	ea.	10	40		40
Man door	1	ea.	10	10		10
Sliding door	1	ea.	10	10		10
Wall fans	2	ea.	10	20		20
Cladding	1	lot	1180	1180		1180
Penthouse bridge crane	1	ea.	240	240		240

9.21 Ductwork to baghouse labor hours sheet 1

Scope	Qty	Unit	MH/Qty	BM	PF	Total MH
Ductwork				**480**		**480**
Vertical duct	6	ea.	20	120		120
Expansion joint	2	ea.	60	120		120
Weld duct at FJ	6	ea.	40	240		240

9.22 Ash tank 15′ dia × 40′ labor hours

	Qty	Unit	MH/Qty	BM	PF	Total MH
Ash tank				**2060**		**2060**
Place floor sheets on foundation	1	ea.	60	60		60
Weld floor sheets	106	lf	1.13	120		3
12 sheets (7.9′W × 8′H) ring/5 rings						
First ring—welded						
Set first ring sheets	12	ea.	5	60		60
Vertical—fit and weld	96	lf	1.25	120		120
Corner weld (inside/outside)	189	lf	1.16	220		220
Set ring 2–5	48	ea.	5	240		240
Bolt sheets together—vertical and horizontal	1707	ea.	0.26	440		439.7
Center pole/roof structural steel	1	ea.	80	80		80
Place roof sheets on structural	1	ea.	60	60		60
Weld floor sheets	106	lf	1.13	120		120
Stairway to tank (grade)—structural (12′ × 14′ × 20′)						
Erect columns	4	ea.	10	40		40
Column field splice	4	ea.	10	40		40
Platforms w/grating/handrail	3	ea.	33.33	100		100
Stairs/handrail	6	ea.	33.33	200		200
Shim, bolt column base plate to fen	4	ea.	15	60		60
Grout column base plates	4	ea.	10	40		40
Install mixer	1	ea.	20	20		20
Unload auger	1	ea.	20	20		20
Caged ladder	40	ea.	0.5	20		20

9.23 Ash piping bid breakdown

	BM	PF	Total MH
Ash piping		**1524**	**1524**
4″ std CS piping		50	50
3″ std CS piping		69	69

(Continued)

(Continued)

	BM	PF	Total MH
3'/5" std CS piping		101	101
Boiler ash assembly		48	48
Reducer assembly		58	58
Economizer, downcomer, and baghouse piping		1198	1198

9.24 Ash piping

9.24.1 Ash piping labor hours sheet 1

Scope	Qty	Unit	MH/ Qty	BM	PF	Total MH
4" std CS piping					**50**	**50**
4" std 150# CS pipe spool assembly handle pipe	16	DIF	0.28		4.5	4.5
4" std 150# CS pipe spool assembly handle pipe	31	DIF	0.07		2.2	2.2
4" std 150# CS pipe spool assembly handle pipe	39	DIF	0.07		2.8	2.8
4" std 150# CS pipe spool assembly handle pipe	68	DIF	0.07		4.7	4.7
4" std 150# CS pipe spool assembly handle pipe	34	DIF	0.07		2.4	2.4
4" std 150# CS pipe spool assembly handle pipe	200	DIF	0.07		14.0	14.0
4" std LR BW 90 deg elbow CS	42	DIF	0.07		2.9	2.9
4" Butt Weld, std CS	32	DI	0.5		16	16
3" std CS piping					**69**	**69**
3" std 150# CS pipe spool assembly	17	DIF	0.07		1.2	1.2
3" std 150# CS pipe spool assembly	61	DIF	0.07		4.3	4.3
3" std 150# CS pipe spool assembly	16	DIF	0.07		1.1	1.1
3" std 150# CS pipe spool assembly	15	DIF	0.07		1.0	1.0
3" std 150# CS pipe spool assembly	20	DIF	0.07		1.4	1.4
3" std 150# CS pipe spool assembly	62	DIF	0.07		4.3	4.3
3" std 150# CS pipe spool assembly	25	DIF	0.07		1.8	1.8
3" std 150# CS pipe spool assembly	32	DIF	0.07		2.2	2.2
3" std 150# CS pipe spool assembly	274	DIF	0.07		19.2	19.2
3" std 150# CS pipe spool assembly	27	DIF	0.07		1.9	1.9
3" std LR BW 90 deg elbow CS	181	DIF	0.07		12.7	12.7
3" butt weld, std CS	24	DI	0.5		12	12
3" std 150# CS pipe spool assembly	31	DIF	0.07		2.2	2.2
3" std 150# CS pipe spool assembly	36	DIF	0.07		2.5	2.5
3" std 150# CS pipe spool assembly	15	DIF	0.07		1.0	1.0
3'/5" std CS piping					**101**	**101**
5" std 150# CS pipe spool assembly	24	DIF	0.07		1.7	1.7
5" std 150# CS pipe spool assembly	27	DIF	0.07		1.9	1.9

(Continued)

(Continued)

Scope	Qty	Unit	MH/ Qty	BM	PF	Total MH
5" std 150# CS pipe spool assembly	29	DIF	0.07		2.0	2.0
5" std 150# CS pipe spool assembly	44	DIF	0.07		3.1	3.1
5" std 150# CS pipe spool assembly	801	DIF	0.07		56.1	56.1
5" std LR BW 90 deg elbow CS	234	DIF	0.07		16.4	16.4
5" butt weld, std CS	40	DI	0.5		20.0	20.0

9.24.2 Ash piping labor hours sheet 2

Scope	Qty	Unit	MH/ Qty	BM	PF	Total MH
Boiler ash assembly					**48.3**	**48.3**
5' std 150# CS boiler ash assembly	65	DIF	0.07		5	5
3' std 150# CS boiler ash assembly	61	DIF	0.07		4.3	4.3
3' std 150# CS boiler ash assembly	14	DIF	0.07		1.0	1.0
3' std 150# CS boiler ash assembly	16	DIF	0.07		1.1	1.1
5' std 150# CS boiler ash assembly	61	DIF	0.07		4.2	4.2
5" std 150# CS elbow	187	DIF	0.07		13.1	13.1
5" Std CS butt weld, std	40	DI	0.5		20.0	20.0
Reducer assembly					**58**	**58**
4' std 150# CS reducer assembly	181	DIF	0.07		12.7	12.7
4' std CS butt weld, std	32	DIF	0.07		2.2	2.2
4' std 150# CS pipe spool assembly	101	DIF	0.07		7.1	7.1
3" std 150# CS pipe spool assembly	49	DIF	0.07		3.4	3.4
3" std 150# CS pipe spool assembly	49	DIF	0.07		3.4	3.4
3" std 150# CS pipe spool assembly	48	DIF	0.07		3.3	3.3
3' std CS air heater ash pot air inlet spool	8	DIF	0.07		0.6	0.6
3" std baghouse ash pot air inlet spool	16	DIF	0.07		1.1	1.1
3' std boiler ash pot air inlet spool	6	DIF	0.07		0.4	0.4
3" std boiler ash pot air inlet spool	18	DIF	0.07		1.3	1.3
3' std blower air outlet spool	58	DIF	0.07		4.1	4.1
3" std LR BW 90 deg elbow CS	89	DIF	0.07		6.3	6.3
3" butt weld, std CS	24	DI	0.5		12.0	12.0

9.24.3 Ash piping labor hours sheet 3

Scope	Qty	Unit	MH/ Qty	BM	PF	Total MH
Economizer, downcomer, and baghouse piping					1198	1198
5" std 150# CS pipe spool assembly	735	DIF	0.07		51.5	51.5
4" std 150# CS pipe spool assembly pipe	168	DIF	0.07		11.8	11.8
3" std 150# CS pipe spool assembly	123	DIF	0.07		8.6	8.6
5" Reducer assembly	21	DIF	0.07		1.4	1.4
5" elbow	15	DIF	0.07		1.0	1.0
5" tee	14	DIF	0.07		1.0	1.0
5" butt weld, std	155	DI	0.5		77.5	77.5
4" elbow	16	DIF	0.07		1.1	1.1
3" elbow	4	DIF	0.07		0.3	0.3
3" tee	8	DIF	0.07		0.6	0.6
3" elbow	7	DIF	0.07		0.5	0.5
3" butt weld, std	93	DI	0.5		46.5	46.5
3" economizer ash pot aid inlet spool	16	DIF	0.07		1.2	1.2
3" ash pot inlet adapter	27	DIF	0.07		1.9	1.9
3" ash pot inlet adapter	22	DIF	0.07		1.5	1.5
3" downcomer assembly	173	DIF	0.07		12.1	12.1
3" ash drop Assembly	188	DIF	0.07		13.2	13.2
3" downcomer assembly, boiler ash drop	99	DIF	0.07		6.9	6.9
3" Ash hopper outlet assembly	223	DIF	0.07		15.6	15.6
3" ash drop wye branch assembly	80	DIF	0.07		5.6	5.6
3" ash pot inlet adaptor	48	DIF	0.07		3.4	3.4
3" baghouse ash pot, noninsulated	201	DIF	0.07		14.1	14.1
3" air heater ash pot, noninsulated	90	DIF	0.07		6.3	6.3
3" ash pot, noninsulated (stainless steel)	302	DIF	0.07		21.1	21.1
12" knife gate	3	ea.	5.4		16.2	16.2
12" pathway	5	ea.	5.4		27.0	27.0
U-bolt	12	ea.	2.00		24.0	24.0
Pipe support	35	ea.	4.00		140.0	140.0
T-bolt clamp	26	ea.	4.00		104.0	104.0
Flex boot	7	ea.	4.00		28.0	28.0
5" 150# boltup	19	ea.	3.25		61.8	61.8
4" 150# boltup	13	ea.	2.60		33.8	33.8
150# boltup	1	ea.	20.00		20.0	20.0
Ash blower	1	ea.	20.00		20.0	20.0
Hydrotest	1	ea.	142		142.0	142.0
Ash hopper screw conveyor	1.75	Ton	80		140.0	140.0
Ash conditioner	1.71	Ton	80		137.1	137.1

9.25 Sand system bid breakdown

	BM	PF	Total MH
Sand system	**3680**		**3680**
Floor sheets	360		360
12 sheets (7.9′W × 8.6′H) ring/6 rings	1600		1600
Conveyors	1720		1720

9.26 Sand system

9.26.1 Sand system labor hours sheet 1

Scope	Qty	Unit	MH/ Qty	BM	PF	Total MH
Sand system						
Sand tank 15.4 Dia × 52′ H				**360**		**360**
Receive, unload, and haul to erection	1	ea.	180	180		180
Place floor sheets on foundation	1	ea.	60	60		60
Weld floor sheets	108	lf	1.1	120		120
12 sheets (7.9′W × 8.6′H) ring/6 rings				**1600**		**1600**
1st Ring—welded						
Set first ring sheets	12	ea.	5	60		60
Vertical—fit and weld	103	lf	1.16	120		120
Corner weld (inside/outside)	####	lf	1.24	240		240
Set ring 2—6	60	ea.	5	300		300
Bolt sheets together—vertical and horizontal	####	ea.	0.26	460		460
Center pole/roof structural steel	1	ea.	80	80		80
Place roof sheets on structural	1	ea.	60	60		60
Weld floor sheets	108	lf	1.11	120		120
Caged ladder	52	lf	0.38	20		20
Decking, 6″ dia pipe w/PS	1	lot	120	120		120
Sand tank 6′ Dia × 10′ (shop fabr) set in field	1	ea.	20	20		20
Conveyors				**1720**		**1720**
Bucket elevator 20″ × 48″ 8—10′ pieces						
Drive head and neck section	1	ea.	180	180		180
Transition and trough section	1	ea.	300	300		300
Tail and trough	1	ea.	300	300		300
Buckets	120	ea.	4	480		480
Vibrating conveyor set/weld down	36.4	ea.	10.44	380		380
Cone connections	3	ea.	20	60		60
Diverter valve	1	ea.	20	20		20

9.27 BOP interconnecting pipe

9.27.1 BOP interconnecting pipe labor hours sheet 1

Scope	Qty	Unit	MH/Qty	BM	PF	Total MH
BOP interconnecting pipe					9180	9180
2" misc. drains and vents	400	lf	1.80		720	720
0.75" vent and drain on economizer	60	lf	1.80		108	108
6" feedwater to economizer	200	lf	2.40		480	480
6" economizer feedwater to boiler	200	lf	2.40		480	480
2" steam to sample station	100	lf	1.80		180	180
2" Spray dryer drains	100	lf	1.80		180	180
2" Instrument air	1600	lf	1.80		2880	2880
2" air to sand reinjection	120	lf	1.80		216	216
2" view port cooling air	200	lf	1.80		360	360
2" atomizing air	200	lf	1.80		360	360
6" natural gas supply	400	lf	2.40		960	960
2" potable water	500	lf	1.80		900	900
6" cooling water	160	lf	2.40		384	384
2" bearing cooling water (supply and return)	240	lf	1.80		432	432
2" condensate return	300	lf	1.80		540	540

9.28 Bid breakdown circulating fluidized bed boiler (FBC)

	BM	Laborer	PF	Total MH
Circulating fluidized bed boiler (FBC)	**36,270**	**40**	**21,944**	**58,254**
Combustor	8902	40		8942
Boiler	5368		5192	10,560
Boiler circulating water			6048	6048
Fans	1160			1160
Fans—FD, ID, OFA ductwork	1940			1940
Economizer/inlet duct/hoppers	1520			1520
Multicione	1440			1440
Spray dryer	3560			3560
Baghouse	6160			6160
Ductwork	480			480
Ash tank	2060			2060
Ash piping			1524	1524
Sand system	3680			3680
BOP interconnecting pipe			9180	9180

Chapter 10

Bid assurance

10.1 Introduction

Accurate, reliable, and quality cost estimates prevent contractors from losing money and the customer from overpaying. The accuracy of a cost estimate relies on:

- quality of the project plan
- level the estimator defines project scope
- accuracy of the cost information
- quality of tools and procedures the estimator uses

The "optimum bid" or the "best bid" will result in a successful bid. When the bid leads to winning a job, then there is an opportunity to verify the estimate's accuracy, reliability, and quality. This chapter describes unbalanced bidding strategy, analysis of estimates, estimate errors, and estimate assurance.

10.2 Unbalanced bidding strategy

For many reasons the management will reduce the bid price and estimators will decrease the unit labor units. The reduction of the total price and decreasing the unit cost on labor is biased estimating and the adjustment is called unbalanced estimating.

Main reasons for unbalanced bidding are:

- front loading—bidding a higher percent than is allowed for mobilization cost,
- using preliminary estimating methods with less information,
- using detail of data warehouses, which has not been verified or matches work scope, and
- total price reduction.

A mathematically unbalanced bid is one containing lump sum or unit bid items that do not reflect reasonable actual cost plus a reasonable proportionate share of the bidder's anticipated profit, overhead cost, which contractor anticipates for the performance of the items in question.

Industrial Construction Estimating Manual. DOI: https://doi.org/10.1016/B978-0-12-823362-7.00010-7

A balanced bid is one where each bid item includes direct cost, overhead, other cost, and markup. Each item has a proportionate share of the cost that is required for the performance of the task.

The unit-quantity model is a balanced unit estimate. The contractor assumes the risks and disadvantages of an unbalanced bid, and the bid can result in losses.

10.3 Analysis of estimates

Historical data cost coded and collected in the field must be verified by statistical analysis and the actual cost compared with the original estimate. Fig. 10.1 illustrates the distribution of estimates plotted against actual cost. The data in Fig. 10.1 has been normalized using regression analysis. Deviations in the cost estimates are expressed by the ratio C_{pe}/C_a. This ratio equals 1 when actual cost equals the estimated cost and the ratio is the *break-even* point. Points to the left of the x axes are negative because the contractor is operating at a loss. Depending on bid strategy and objectives, contractor will profit to the right of the break-even point. The contractor's survival depends on balanced unit estimates.

10.3.1 Regression analysis—deviation of estimate ratio from actual value

C_{pe}/C_a, x	No. of estimates, y
0	−100
1	0
2	100
3	200
4	300

COVAR (R1, R2)	200.00
VARP (R2)	20,000.00
Slope (R1, R2)	100.0000
Intercept (R1, R2)	−100.0000

10.3.2 Estimate error

Accurate construction estimating is essential for contractors to prosper. The deviation between the estimate and actual cost for losing estimates may be due to errors in the estimate. Errors are not obvious immediately particularly when small. Errors can cost time, money, and reputation.

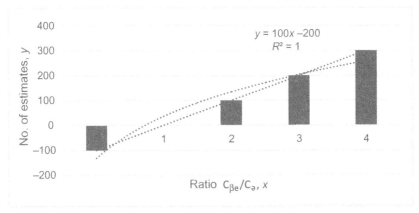

FIGURE 10.1 Distribution of estimates plotted against actual cost.

Three kinds of errors in estimating are:

- mistakes
- policy
- risk

10.3.3 Mistakes

- Underestimating labor—underestimating time a task will take
- Last-minute changes—forgotten line items
- Under estimating margin
- Selective bidding (union vs nonunion) and bidding too many projects
- Error in takeoff off scope
- Spreadsheets with too much detail
- Arithmetic, and error in formulas
- Risk—lack of or improper factoring
- Poorly defined scope of work—not understanding complete scope of work

10.3.4 Policy

- Excessive or low overhead ratios
- Using data that has not been verified
- Estimating with data sets or books using unconfirmed data
- Data collection

10.3.5 Risk

- Bid is intentionally low to keep workforce busy.
- Risk error is the difference between contractors' bid and the winning bid.

- Material and subcontractor pricing.
- Incomplete clarifications and exceptions.

An important objective of the contractor is to prepare exact estimates. However, estimates will deviate from the actual cost and reasons for the deviation between the estimate and actual cost include pitfalls that negatively impact the estimate and undermine the accuracy and validity of the estimate.

Estimate pitfalls:

- Errors and unbalanced bidding.
- Evaluation of actual value between the estimate and historical data is not timely.
- Historical data is not collected to allow a comparison.
- There is no analysis of low and unsuccessful bids.
- Historical data is not verified.
- Too much time and effort spent preparing the estimate.

Estimator pitfalls

- Padding estimator adds a factor to work task, a cushion, that increases cost.
- Estimator uses third party data that is misleading relaying on data without understanding how data was derived.
- Estimator does not read the project documents carefully.
- Estimator does not factor for labor productivity and risk.
- Estimator does not divide task between multiple resources.
- Estimator does not update cost estimate and scope changes.
- Estimator lacks experience with similar projects.
- Estimator does not incorporate cost or enter cost correctly.
- Estimator uses material takeoff spreadsheets with too much detail that causes mistakes in data entry and errors.

Mistakes, errors, and estimator errors can cause estimates to lose money and cause the contractor to lose business. Errors are prevented by planning, well-thought-out policies, data collection, and estimate analysis.

10.4 Estimate assurance

Comparison of actual-to-estimate values is important in the cost of construction work. Detailed estimating using the unit-quantity model is significant in bid assurance. The unit-quantity model is a balanced bid strategy. Standard unit of construction work is completed before construction is started. In estimating construction work the contractor must determine the materials that are required to complete the work, time and labor productivity for each task, and schedule indirect and staff for complete total construction. The aim is to specify the most effective method to complete the construction at the lowest practical cost. This planning occurs when the unit-quantity model is used to estimate the construction work. Unit cost estimated simply associates unit cost with each assembly involved in a construction process. The estimate is accurate and reliable, using comparable historical data (assemblies used previously) with evidence to justify the unit cost for each assembly. Detail estimating influences database development, computer advantages, is consistent and has many advantages compared to other estimating methods.

Chapter 11

Detailed estimating applications to construction

11.1 Introduction

Contractors regardless of their size have difficulty estimating and focusing on the best and most profitable work. The contractor must avoid bidding too low, losing money, and cannot bid too high, and not win new work. The estimate needs to be high enough to make a profit but low enough to win a job. With all the pitfalls that undermine the accuracy and validity of an estimate, the contractor must have estimates that are accurate and consistent to make a profit and grow their reputation. The focus of this chapter is how to overcome and sustain profitability and streamline the estimating process.

This chapter includes practical examples of applications to enable the reader to use the *industrial construction estimating process* to prepare process piping, equipment and civil estimates to set up databases.

The following illustrative example for the *industrial construction estimating process* enables the reader to streamline the estimating process using the detailed estimating method to:

- evaluate the accuracy and verify the historical data collected in the field for process piping, equipment and civil work installed in a combined cycle power plant
- provides the unit-quantity model for detailed estimating that has many advantages compared to other estimating methods.

Industrial construction estimating process

- Scope of work is defined by the plant process, piping and instrument diagrams (P&IDs) and process flow diagrams (PFDs), and field-specific scope.
- Make detailed material takeoff (MTO) for piping.
- Set up cost codes—job cost by cost code and type.
- Collect direct craft data in the field; field report.
- Summarize, verify, and validate data using the regression model.
- Set up databases for industrial process plants.
- Apply man-hour units to the quantified MTO.

Industrial Construction Estimating Manual. DOI: https://doi.org/10.1016/B978-0-12-823362-7.00011-9

- Enter the man-hour rates from the tables and takeoff quantities into the estimate sheets and calculate the direct craft labor using the unit-quantity model (man-hour rate × quantity) and then estimated man-hours are compared to actual.

11.2 Illustrative example for construction estimating process—lube oil supply

11.2.1 Scope of work is defined by the plant process, piping and instrument diagrams, and process flow diagrams

Lube oil supply

20 ft. 6″ s10s 304L SS pipe
20 ft. 4″ s10s 304L SS pipe
40 ft. 2″ s40s 304L SS pipe
20 ft. 1″ s40s 304L SS pipe
80 ft. 4″ s10s 304L SS pipe
20 ft. 2″ s40s 304L SS pipe
20 ft. 2″ s40s 304L SS pipe
40 ft. 4″ s10s 304L SS pipe
40 ft. 2″ s40s 304L SS pipe

11.2.2 Make detailed material takeoff for lube oil supply piping

2.7.1 Steam Turbine
Generator (STG)
vendor piping sheet 1

Line	Material	Size	Sch/ Thk	Pipe	BW, SW	Valve	Boltup	Instrument	PS
Lube oil supply									
Pipe-304L SS	304L SS	6	S10S	20	4		1		
Pipe-304L SS	304L SS	4	S10S	20	7		1		1
Pipe-304L SS	304L SS	2	S40S	40	13		1		1
Pipe-304L SS	304L SS	1	S40S	20	17		2		2
Pipe-304L SS	304L SS	4	S10S	80	21		1		4
Pipe-304L SS	304L SS	2	S40S	20	9		1		
Pipe-304L SS	304L SS	2	S40S	20	9		1		
Pipe-304L SS	304L SS	4	S10S	40	13	1	1		2
Pipe-304L SS	304L SS	2	S40S	40	13		1		4

11.2.3 Set up cost codes—job cost by cost code and type

Job cost by cost code and type—lube oil supply piping

Cost code	Type	Qty
xxxxxx	6" s10s 304L SS pipe	20
xxxxxx	4" s10s 304L SS pipe	20
xxxxxx	2" s40s 304L SS pipe	40
xxxxxx	1" s40s 304L SS pipe	20
xxxxxx	4" s10s 304L SS pipe	80
xxxxxx	2" s40s 304L SS pipe	20
xxxxxx	2" s40s 304L SS pipe	20
xxxxxx	4" s10s 304L SS pipe	40
xxxxxx	2" s40s 304L SS pipe	40

11.2.4 Collect direct craft data in the field—field report

Foreman's report
Project: Combined cycle power plant STG vendor LO piping

Foreman	Date		
Craft: pipefitter			
Cost code	Type	Qty	MH
xxxxxx	6" s10s 304L SS pipe	20	37
xxxxxx	4" s10s 304L SS pipe	20	36
xxxxxx	2" s40s 304L SS pipe	40	51
xxxxxx	1" s40s 304L SS pipe	20	24
xxxxxx	4" s10s 304L SS pipe	80	119
xxxxxx	2" s40s 304L SS pipe	20	32
xxxxxx	2" s40s 304L SS pipe	20	32
xxxxxx	4" s10s 304L SS pipe	40	69
xxxxxx	2" s40s 304L SS pipe	40	58

11.2.5 Summarize, verify, and validate data using the regression model

The historical data from the field report is verified using the regression model.

Least squares regression model
Facility—Combined cycle power plant
Data for input: Man-hours for field installation lube oil supply piping
Quantity (x): R1 = 20, 20, 40, 20, 80, 20, 20, 40, 40
Man-hour (y): R2 = 37, 36, 51, 24, 119, 32, 32, 69, 58
Table linear regression: Fitting a straight line (Fig. 11.1).

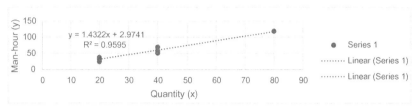

FIGURE 11.1 Installation lube oil supply piping.

X	y
Qty	MH
x	y
20	37
20	36
40	51
20	24
80	119
20	32
20	32
40	69
40	58
COVAR (R1, R2)	509.23
VARP (R2)	760.12
SLOPE (R1, R2)	1.4322
INTERCEPT (R1, R2)	2.9741

CORREL (R1, R2) = correlation coefficient	0.9795
CORREL (R1, R2)^2 = coefficient determination	0.9595

The coefficient of determination is $R^2 = 0.9595$ and the correlation coefficient, $R = 0.9795$, is a strong indication of correlation. A percentage of 98.5 of the total variation on Y can be explained by the linear relationship between X and Y (described by the regression equation; $Y = 1.4322x + 2.9741$). The relationship between X and Y variables is such that as X increases, Y also increases.

11.3 Man-hour database for combined cycle power plant and industrial plant

11.3.1 Schedule A—Combined cycle power plant piping

Standard labor-estimating units

Facility—Combined cycle power plant	Large bore piping	Small bore piping
	Unit of measure	Unit of measure
Description	Man-hours per unit	Man-hours per unit
Handle and install pipe, carbon steel, welded joint	Diameter inch feet	MH/LF
WT ≤ 0.375″	0.07	0.18
0.406″ ≤ WT ≤ 0.500″	0.09	0.23
0.562″ ≤ WT ≤ 0.688″	0.11	0.28
0.718″ ≤ WT ≤ 0.938″	0.14	0.35
1.031″ ≤ WT ≤ 1.219″	0.2	0.5
1.250″ ≤ WT ≤ 1.312″	0.25	0.75
Welding butt welds, carbon steel, arc-uphill	Diameter inch	MH/EA
WT ≤ 0.375″	0.5	1.1
0.406″ ≥ WT ≤ 0.500″	0.55	1.2
0.562″ ≥ WT ≤ 0.688″	1.05	2.2
0.718″ ≥ WT ≤ 0.938″ (PWHT)	1.2	2.45
1.031″ ≥ WT ≤ 1.219″ (PWHT)	1.45	2.7
1.250″ ≤ WT ≤ 1.312″ (PWHT)	2.2	4.4
Olet- (SOL, TOL, WOL)	2 × BW	2 × BW
Stub in	1.5 × BW	1.5 × BW
Socketweld		Per SW table
PWHT craft support labor	0.45	1
Boltup of flanged joints by weight class	Diameter inch	MH/EA
150/300 boltup	0.4	1
600/900 boltup	0.5	1.2
1500/2500 boltup	0.65	1.6
Handle valves by weight class	Diameter inch	MH/EA
150 and 300 manual valves	0.45	1
600 and 900 manual valves	0.9	1.8
Heavier manual valve ≥ 1500	1.8	2

Facility—Industrial plant

11.3.2   Hydrostatic testing

	Man-hours per?lineal?foot				
	Wall?thickness in inches				
Pipe?size	0.375″?or?less	0.406″−0.500″	0.562″−0.688″	0.718″−0.938″	1.031″−1.219″
0.5	0.022	0.028	0.034	0.042	0.060
0.75	0.022	0.028	0.034	0.042	0.060

(Continued)

(Continued)

	Man-hours per?lineal?foot				
	Wall?thickness in inches				
1	0.022	0.028	0.034	0.042	0.060
1.5	0.022	0.028	0.034	0.042	0.060
2	0.022	0.028	0.034	0.042	0.060
2.5	0.021	0.027	0.033	0.042	0.060
3	0.025	0.032	0.040	0.050	0.072
4	0.034	0.043	0.053	0.067	0.096
6	0.050	0.065	0.079	0.101	0.144
8	0.067	0.086	0.106	0.134	0.192
10	0.084	0.108	0.132	0.168	0.240
12	0.101	0.130	0.158	0.202	0.288
14	0.118	0.151	0.185	0.235	0.336
16	0.134	0.173	0.211	0.269	0.384
18	0.151	0.194	0.238	0.302	0.432
20	0.168	0.216	0.264	0.336	0.480
24	0.202	0.259	0.317	0.403	0.576

Man-hours to place/remove blinds, open/close valves, removal/replacement of valves and specialty item + s and pipe sections as required, and drain lines after testing

11.3.3 Schedule G—Alloy and nonferrous weld factors

Welding percentages for alloy and nonferrous metals	Material classification- group numbers							
Pipe size	1	2	3	4	5	6	7	8
2	0.25	0.54	0.20	0.58	2.11	2.25	0.225	0.45
3	0.275	0.58	0.23	0.61	2.15	2.32	0.25	0.495
4	0.30	0.61	0.25	0.68	2.22	2.35	0.28	0.54
5	0.315	0.63						0.57
6	0.345	0.65	0.30	0.75	2.28	2.40	0.30	0.62
8	0.39	0.74	0.50	0.88	2.38	2.50	0.34	0.70
10	0.425	0.85	0.75	0.95	2.45	2.75	0.375	0.765
12	0.45	2.00	0.80	2.04	2.50	3.00	0.40	0.81
14	0.49	2.15						0.88
16	0.525	2.23						0.945
18	0.59	2.30						2.06
20	0.65	2.45						2.17
24	0.73							2.24

GROUP 1–CHROME MOLYBLENUM STEEL
    CHROME —1/2%–13%
    MOLY—to 1%
GROUP 2—18–8 STAINLESS STEEL
    TY, 304, 316, 347
GROUP 3—COPPER, BRASS, EVERDUR
GROUP 4—ALUMINUM, MONEL and COPPER
    CHROME–NICKEL
GROUP 5—NICKEL
GROUP 6—HASTELLOY
GROUP 7—GALVANIZED
GROUP 8—A335–P91

11.4 Lube oil supply piping estimate

- Apply man-hour units to the quantified MTO.
- Enter the man-hour rates from the tables and takeoff quantities into the estimate sheets and calculate the direct craft labor using the unit-quantity model (man-hour rate × quantity) then estimated man-hours are compared to actual.

Facility—Combined cycle power plant
STG vendor lube oil supply piping

11.4.1 Estimate sheet 1-Handle and install pipe-welded joint

| Description | Size | Pipe handle | | | PF | Hydro | | PF | Total |
		LF	DIF	MH/ DIF	MH	DIF	MH/ DIF	MH	MH
Lube oil supply									
Pipe-304L SS	6	20	120	0.07	8.4	120	0.050	6	14.4
Pipe-304L SS	4	20	80	0.07	5.6	80	0.034	2.72	8.3
Pipe-304L SS	2	40	80	0.18	14.4	80	0.022	1.76	16.2
Pipe-304L SS	1	20	20	0.18	3.6	20	0.022	0.44	4.0
Pipe-304L SS	4	80	320	0.07	22.4	320	0.034	10.88	33.3
Pipe-304L SS	2	20	40	0.18	7.2	40	0.034	1.36	8.6
Pipe-304L SS	2	20	40	0.18	7.2	40	0.034	1.36	8.6
Pipe-304L SS	4	40	160	0.07	11.2	160	0.034	5.44	16.6
Pipe-304L SS	2	40	80	0.18	14.4	80	0.034	2.72	17.1
Column totals		300	940		94.4	940		32.7	127.1

Facility—Combined cycle power plant
STG vendor lube oil supply piping

11.4.2 Estimate sheet 2-Welding: BW, SW, PWHT arc-uphill

Description	Size	BW JT	SW JT	DI	BW MH/ DI	SW MH/ JT	Factor	PWHT MH/ JT	PF MH
Lube oil supply									
Pipe-304L SS	6	4	0	24	0.50	0	1.65	0	19.8
Pipe-304L SS	4	7	0	28	0.50	0	1.61	0	22.5
Pipe-304L SS	2	13	0	26	0.80	0	1.54	0	32.0
Pipe-304L SS	1	17	0	17	0.65	0	1.54	0	17.0
Pipe-304L SS	4	21	0	84	0.50	0	1.61	0	67.6
Pipe-304L SS	2	9	0	18	0.80	0	1.54	0	22.2
Pipe-304L SS	2	9	0	18	0.80	0	1.54	0	22.2
Pipe-304L SS	4	13	0	52	0.50	0	1.61	0	41.9
Pipe-304L SS	2	13	0	26	0.80	0	1.54	0	32.0
Column totals		106	0	293					277.3

Facility—Combined cycle power plant
STG vendor lube oil supply piping

11.4.3 Estimate sheet 3-Boltup of flanged joint by weight class

Description	Size	150/ 300 Boltup	600/ 900 Boltup	1500/ 2500 Boltup	DI	MH/ DI	MH/ DI	MH/ DI	PF MH
Lube oil supply									
Pipe-304L SS	6	1	0	0	6	0.4	0.5	0.65	2.4
Pipe-304L SS	4	1	0	0	4	0.4	0.5	0.65	1.6
Pipe-304L SS	2	1	0	0	2	0.4	0.5	0.65	0.8
Pipe-304L SS	1	2	0	0	2	0.4	0.5	0.65	0.8
Pipe-304L SS	4	1	0	0	4	0.4	0.5	0.65	1.6
Pipe-304L SS	2	1	0	0	2	0.4	0.5	0.65	0.8
Pipe-304L SS	2	1	0	0	2	0.4	0.5	0.65	0.8
Pipe-304L SS	4	1	0	0	4	0.4	0.5	0.65	1.6
Pipe-304L SS	2	1	0	0	2	0.4	0.5	0.65	0.8
Column totals		10			28				11.2

Facility—Combined cycle power plant
STG vendor lube oil supply piping

11.4.4   Estimate sheet 4-Handle valves by weight class

Description		150/ 300	600/ 900	1500/ 2500					PF
Lube oil supply	Size	Valve	Valve	Valve	DI	MH/ DI	MH/ DI	MH/ DI	MH
Pipe-304L SS	6								
Pipe-304L SS	4	0	0	0	0.45	0.9	1.8	0	
Pipe-304L SS	2	0	0	0	0.45	0.9	1.8	0	
Pipe-304L SS	1	0	0	0	0.45	0.9	1.8	0	
Pipe-304L SS	4	0	0	0	0.45	0.9	1.8	0	
Pipe-304L SS	2	0	0	0	0.45	0.9	1.8	0	
Pipe-304L SS	2	0	0	0	0.45	0.9	1.8	0	
Pipe-304L SS	4	0	0	0	0.45	0.9	1.8	0	
Pipe-304L SS	2	1	0	0	2	0.45	0.9	1.8	0.9
	3		0	0	0	0.45	0.9	1.8	0
Column totals		1			2				0.9

Facility—Combined cycle power plant
STG vendor lube oil supply piping

11.4.5   Estimate sheet 5-Pipe supports

				Pipe			PF
Description	Material	Size	Sch/ Thk	Support	DI	MH/ DI	MH
Lube oil supply							
Pipe-304L SS	304L SS	6	S10S		0	1	0
Pipe-304L SS	304L SS	4	S10S	1	4	1	4
Pipe-304L SS	304L SS	2	S40S	1	2	1	2
Pipe-304L SS	304L SS	1	S40S	2	2	1	2
Pipe-304L SS	304L SS	4	S10S	4	16	1	16
Pipe-304L SS	304L SS	2	S40S		0	1	0
Pipe-304L SS	304L SS	2	S40S		0	1	0
Pipe-304L SS	304L SS	4	S10S	2	8	1	8
Pipe-304L SS	304L SS	2	S40S	4	8	1	8
Column totals				14	40		40

Facility—Combined cycle power plant
STG vendor lube oil supply piping

11.4.6 Estimate sheet 6-Instrument

						PF
Description	Material	Size	Sch/The	Instrument	MH/EA	MH
Lube oil supply						
Pipe-304L SS	304L SS	6	S10S	0	1.2	0
Pipe-304L SS	304L SS	4	S10S	0	1.2	0
Pipe-304L SS	304L SS	2	S40S	0	1.2	0
Pipe-304L SS	304L SS	1	S40S	0	1.2	0
Pipe-304L SS	304L SS	4	S10S	0	1.2	0
Pipe-304L SS	304L SS	2	S40S	0	1.2	0
Pipe-304L SS	304L SS	2	S40S	0	1.2	0
Pipe-304L SS	304L SS	4	S10S	0	1.2	0
Pipe-304L SS	304L SS	2	S40S	0	1.2	0
				0		
Column totals				0		0

Facility—Combined cycle power plant
STG vendor lube oil supply piping

11.4.7 Estimate sheet 7-Summary HP piping and supports

	PF	MH/LF
Description	MH	
Estimate sheet 1-handle and install pipe-welded joint	127.1	
Estimate sheet 2-welding: BW, SW, PWHT arc-uphill	277.3	
Estimate sheet 3-boltup of flanged joint by weight class	11.2	
Estimate sheet 4-handle valves by weight class	0.9	
Estimate sheet 5-pipe supports	40	
Estimate sheet 6-instrument	0	
Column totals	456	1.52

11.5 Piping summary converted to MH/LF

Lube oil supply	Pipe	Pipe Hdl	Welding	Boltup	Valve	PS	Instrument	MH	MH/ LF
Pipe-304L SS	20	14.4	19.8	2.4	0	0	0	36.6	1.83
Pipe-304L SS	20	8.3	22.5	1.6	0	4	0	36.5	1.82
Pipe-304L SS	40	16.2	32.0	0.8	0	2	0	51.0	1.27
Pipe-304L SS	20	4.0	17.0	0.8	0	2	0	23.9	1.19
Pipe-304L SS	80	33.3	67.6	1.6	0	16	0	118.5	1.48
Pipe-304L SS	20	8.6	22.2	0.8	0	0	0	31.5	1.58

(Continued)

(Continued)

Lube oil supply	Pipe	Pipe Hdl	Welding	Boltup	Valve	PS	Instrument	MH	MH/ LF
Pipe-304L SS	20	8.6	22.2	0.8	0	0	0	31.5	1.58
Pipe-304L SS	40	16.6	41.9	1.6	0.9	8	0	69.0	1.73
Pipe-304L SS	40	17.1	32.0	0.8	0	8	0	58.0	1.45
Column totals	300	127	277	11	1	40	0	456	1.52

11.6   Excel estimate sheet

Lube oil supply	Material	SIZE	Sch	MH/LF	Qty	lf	MH
Pipe-304L SS	304L SS	6	S10S	1.83	20	lf	37
Pipe-304L SS	304L SS	4	S10S	1.82	20	lf	36
Pipe-304L SS	304L SS	2	S40S	1.27	40	lf	51
Pipe-304L SS	304L SS	1	S40S	1.19	20	lf	24
Pipe-304L SS	304L SS	4	S10S	1.48	80	lf	119
Pipe-304L SS	304L SS	2	S40S	1.58	20	lf	32
Pipe-304L SS	304L SS	2	S40S	1.58	20	lf	32
Pipe-304L SS	304L SS	4	S10S	1.73	40	lf	69
Pipe-304L SS	304L SS	2	S40S	1.45	40	lf	58

11.7   STG-lube oil supply piping installation man-hours

Facility—Combined cycle power plant	Actual	Estimated
Description	MH	PF
Lube oil supply	456	456
Pipe-304L SS	37	37
Pipe-304L SS	36	36
Pipe-304L SS	51	51
Pipe-304L SS	24	24
Pipe-304L SS	119	119
Pipe-304L SS	32	32
Pipe-304L SS	32	32
Pipe-304L SS	69	69
Pipe-304L SS	58	58

11.8   Illustrative example to develop a database for tank farm boltup flanged joints

The historical data is collected on site every day and summarized in a spreadsheet. The data for bolt up is cost-coded and collected in the field.

11.8.1 Field data report

The field report is collected in the field for similar work, and a spreadsheet is devised for the data. The report is used for time control and to find the number of man-hours for a task. The spreadsheet prepares the data for statistical analysis. The estimator determines the productivity rate, and the rate is used for future cost analysis and estimating similar scopes of work.

Foreman	Date	
Craft: PF		
Cost code	Type	MH/JT
xxxxxx	2 150/300 boltup	1.0
xxxxxx	2.5 150/300 boltup	1.0
xxxxxx	3 150/300 boltup	1.2
xxxxxx	4 150/300 boltup	1.6
xxxxxx	6 150/300 boltup	2.4
xxxxxx	8 150/300 boltup	3.2
xxxxxx	10 150/300 boltup	4.0
xxxxxx	12 150/300 boltup	4.8
xxxxxx	14 150/300 boltup	5.6
xxxxxx	16 150/300 boltup	6.4
xxxxxx	18 150/300 boltup	7.2
xxxxxx	20 150/300 boltup	8.0
xxxxxx	24 150/300 boltup	9.6
xxxxxx	2 600 boltup	1.3
xxxxxx	2.5 600 boltup	1.5
xxxxxx	3 600 boltup	1.8
xxxxxx	4 600 boltup	2.4
xxxxxx	6 600 boltup	3.6
xxxxxx	8 600 boltup	4.8
xxxxxx	10 600 boltup	6.0
xxxxxx	12 600 boltup	7.2
xxxxxx	14 600 boltup	8.4
xxxxxx	16 600 boltup	9.6
xxxxxx	18 600 boltup	10.8
xxxxxx	20 600 boltup	12.0
xxxxxx	24 600 boltup	14.4

The historical data from the field report is verified using the regression model.

11.8.2 Least squares regression model—150 lb/300 lb

Facility—Petroleum—tank farm piping
  Data for input: Man-hours for field 150 lb/300 lb boltup
  Pipe size (x): R1 = 2, 2.5. 3, 4, 6, 8, 10, 12, 14, 16 18, 20, 24
  MH/JT (y): R2 = 1.0, 1.0, 1.2, 1.6, 2.4, 3, 2, 4.0, 4.8, 5.6, 6.4, 7.2, 8.0, 9.6
  Table linear regression: Fitting a straight line (Fig. 11.2).

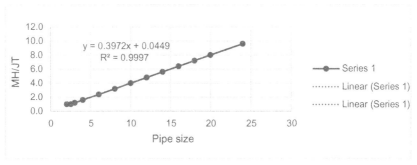

FIGURE 11.2 Tank farm field 150 lb/300 lb boltup.

X	Y
Pipe size	MH/JT
X	Y
2	1.0
2.5	1.0
3	1.2
4	1.6
6	2.4
8	3.2
10	4.0
12	4.8
14	5.6
16	6.4
18	7.2
20	8.0
24	9.6
COVAR (R1, R2)	19.38
VARP (R2)	7.70
SLOPE (R1, R2)	0.3972
INTERCEPT (R1, R2)	0.0449

CORREL (R1, R2) = correlation coefficient	− 1.00
CORREL (R1, R2)^2 = coefficient determination	1.00

The coefficient of determination, R^2, is exactly $+1$ and indicates a positive fit. All data points lie exactly on the straight line. The relationship between X and Y variables is such that as X increases, Y also decreases.

Field data is verified for field 150 lb/300 lb boltup

11.8.3 Least squares regression model—600 lb field boltup

Facility—Petroleum—tank farm piping
  Data for input: Man-hours for field 600 lb bolt
  Pipe size (x): R1 = 2, 2.5. 3, 4, 6, 8, 10, 12, 14, 16 18, 20, 24
  MH/JT (y): R2 = 1.3, 1.5, 1.8, 2.4, 3.6, 4.8, 6.0, 7, 2, 8.4, 9.6, 10.8, 12.0, 14.4
  Table linear regression: fitting a straight line (Fig. 11.3).

x	y
Pipe size	MH/JT
2	1.3
2.5	1.5
3	1.8
4	2.4
6	3.6
8	4.8
10	6.0
12	7.2
14	8.4
16	9.6
18	10.8
20	12.0
24	14.4
COVAR (R1, R2)	29.21
VARP (R2)	17.49
SLOPE (R1, R2)	0.5986
INTERCEPT (R1, R2)	0.0225

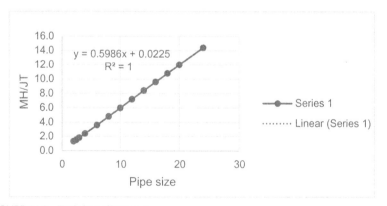

FIGURE 11.3 Tank farm 600 lb boltup.

CORREL (R1, R2) = correlation coefficient	− 1.00
CORREL (R1, R2)^2 = coefficient determination	1.00

The coefficient of determination, R^2, is exactly $+1$ and indicates a positive fit. All data points lie exactly on the straight line. The relationship between X and Y variables is such that as X increases, Y also decreases.

Field data is verified for field 600 lb boltup.

11.8.4   Verification of tank farm–bolted flange connections

- man-hours for field 150 lb/300 lb boltup
- man-hours for field 600 lb bolt

$R^2 = 1$ indicates that the model fits the data.

The historical data collected, summarized, and analyzed using the regression model is verified, and the estimator sets up the unit man-hour table, for tank farm boltup, to include in the piping database.

Table tank farm field boltup MH/JT

Facility—Petroleum—tank farm piping man-hours per joint

Pipe	Pressure rating	
Size	150 lb/300 lb	600 lb
2	1.0	1.3
2.5	1.0	1.5
3	1.2	1.8
4	1.6	2.4
6	2.4	3.6
8	3.2	4.8
10	4.0	6.0
12	4.8	7.2
14	5.6	8.4
16	6.4	9.6
18	7.2	10.8
20	8.0	12.0
24	9.6	14.4

Man-hours per joint to bolt up valves, expansion joints, flanged fittings and spools.

11.9 Illustrative example for verification of LP piping and supports—ASME Section 1 installation

11.9.1 LP piping and supports MTO

				LB			LB	LB		
Line	Material	Size	Sch/ The	Pipe	BW	PWHT	Valve	Boltup	Instrument	PS
LP piping and supports—ASME Section 1										
LP-10 steam drum to LP SHTR 1	SA-106-B	14	std	26					4	
LP-10 steam drum to LP SHTR 1	SA-106-B	10	40	34	4					
LP-10 steam drum to LP SHTR 1	SA-106-B	8	40	12	3					
LP-14 SHTR 1 to LP SHTR 2	SA-106-B	10	40	32	2				2	
LP-15 SHTR 1 to LP SHTR 2	SA-106-B	10	40	30	2				2	

11.9.2 LP piping and supports—ASME Section 1 field estimate

				MH	Qty	Unit	Qty	PF
Line	Material	Size	Sch/ Thk	MH_a	n_i		n_i	$n_i\,MH_a$
LP piping and supports— ASME Section 1					134.0	lf		300.2
LP-10 steam drum to LP SHTR 1	SA-106-B	14	std	2.22	26	lf	26	57.7
LP-10 steam drum to LP SHTR 1	SA-106-B	10	40	2.32	34	lf	34	78.8
LP-10 steam drum to LP SHTR 1	SA-106-B	8	40	3.11	12	lf	12	37.3
LP-14 SHTR 1 to LP SHTR 2	SA-106-B	10	40	2.01	32	lf	32	64.4
LP-15 SHTR 1 to LP SHTR 2	SA-106-B	10	40	2.07	30	lf	30	62.0

11.9.3 Least squares regression model LP piping

Facility—Combined cycle power plant
 Data for input: Man-hours for LP piping and supports
 Quantity (x): R1 = 26, 34, 12, 32, 30
 MM/LF (y): R2 = 57.7, 78.8, 37.3, 64.4, 62.0
 Table linear regression: Fitting a straight line (Fig. 11.4).

x	Y
Qty (lf)	MH/LF
26	57.7
34	78.8
12	37.3
32	64.4
30	62.0
COVAR (R1, R2)	100.43
VARP (R2)	179.31
SLOPE (R1, R2)	1.6262
INTERCEPT (R1, R2)	16.4588

CORREL (R1, R2) = correlation coefficient	0.9544
CORREL (R1, R2)^2 = coefficient determination	0.9108

The coefficient of determination is $R^2 = 0.9108$ and the correlation coefficient, $R = 0.9544$, is a strong indication of correlation. A percentage of 95.4 of the total variation on Y can be explained by the linear relationship between X and Y (described by the regression equation; $Y = 1.6262x + 16.459$). The relationship between X and Y variables is such that as X increases, Y also increases.

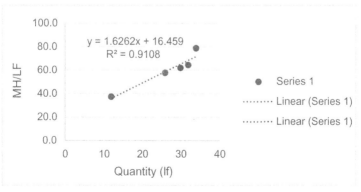

FIGURE 11.4 MH for LP piping and supports.

11.10 Illustrative example for verification of STG utility bridge steel

11.10.1 Power plant structural steel quantity MTO

Project: combined cycle power plant	Revision	X	Date	x/xx/xxxx
			Steel weight	Total steel weight
Equipment description	Component description	Quantity	Tons	Tons
STG utility bridge steel				187
	Light—0–19 lb/ft.	lot	7.6	8.4
	Medium—20–39 lb/ft.	lot	25.4	27.9
	Heavy—40–79 lb./ft.	lot	90	99
	X heavy—80–120 lb/ft.	lot	47	51.7

11.11 STG utility bridge steel field estimate

Description	MH	Qty	Unit	Qty		
	MH_a	n_i		n_i	$n_i\,MH_a$	IW
STG utility bridge steel				187	3453	3453	3453
Light—0–19 lb/ft.	24	8.4	ton	8.4	201.6	201.6	202
Medium—20–39 lb/ft.	23	27.9	ton	27.9	641.7	641.7	642
Heavy—40–79 lb/ft.	18	99	ton	99	1782	1782	1782
X heavy—80–120 lb/ft.	16	51.7	ton	51.7	827.2	827.2	827

11.11.1 Least squares regression model STG utility bridge steel

Facility—Combined cycle power plant
  Data for input: Man-hours for STG utility bridge steel
  Quantity (x): R1 = 8.4, 27.9, 99, 51.7
  MH/ton (y): R2 = 201.6, 641.7, 1782, 827.2
  Table linear regression: Fitting a straight line (Fig. 11.5).

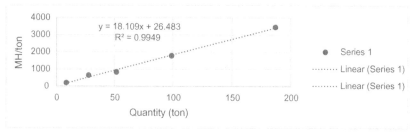

FIGURE 11.5 MH STG utility bridge steel.

X	y
Qty (ton)	MH/ton
187	3453
8.4	201.6
27.9	641.7
99	1782
51.7	827.2
COVAR (R1, R2)	73,580.92
VARP (R2)	1,339,231.31
SLOPE (R1, R2)	18.1085
INTERCEPT (R1, R2)	26.4828

CORREL (R1, R2) = correlation coefficient	0.9975
CORREL (R1, R2)^2 = coefficient determination	0.9949

The coefficient of determination is $R^2 = 0.9949$ and the correlation coefficient, $R = 0.9975$, is a strong indication of correlation. A percentage of 99.7 of the total variation on Y can be explained by the linear relationship between X and Y (described by the regression equation; $Y = 18.109x + 26.483$). The relationship between X and Y variables is such that as X increases, Y also increases.

11.12   Illustrative example for verification of hydrogen plant foundations

11.12.1   MTO hydrogen plant foundation work

Description	QTY	Unit
Foundation slabs—4000PSI	9655	CY
Footing concrete—4000PSI	2416	CY
Pier concrete—4000PSI	275	CY
Elevated floor slab concrete	109	CY
Concrete walls THK 8″–10″	185	CY

11.12.2 Job cost by cost code and type—hydrogen plant foundations

Cost code	Type	Qty	Unit
xxxxxx	Foundation slabs—4000PSI	9655	CY
xxxxxx	Footing concrete—4000PSI	2416	CY
xxxxxx	Pier concrete—4000PSI	275	CY
xxxxxx	Elevated floor slab concrete	109	CY
xxxxxx	Concrete walls THK 8″–10″	185	CY

11.12.3 Civil databases for hydrogen plant

Concrete works	MH	Unit
Foundation slabs—4000PSI	1.831	CY
Footing concrete—4000PSI	2.31	CY
Pier concrete—4000PSI	2.751	CY
Elevated floor slab concrete	3.85	CY
Concrete walls THK 8″–10″	3.852	CY

11.12.4 Excel estimate spreadsheet for hydrogen plant foundations

Description	Historical			Estimate				
	MH	Qty	Unit	Qty		Carpenter	Labor	IW
	MH_a	n_i		n_i	$n_i MH_a$			
Foundation slabs—4000PSI	1.831	9655	CY	9655	17,678		17,678	
Footing concrete—4000PSI	2.31	2416	CY	2416	5582		5582	
Pier concrete—4000PSI	2.751	275	CY	275	758		758	
Elevated floor slab concrete	3.85	109	CY	109	418		418	
concrete walls THK 8″–10″	3.852	185	CY	185	712		712	

11.12.5 Least squares regression model hydrogen plant foundations

Facility—Hydrogen plant
   Data for input: Man-hours for hydrogen plant foundations
   Quantity (x): R1 = 9655, 2416, 275, 109, 185
   Man-hour (y): R2 = 17678, 5582, 758, 418, 712
   Table linear regression: Fitting a straight line (Fig. 11.6).

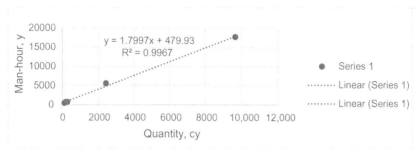

FIGURE 11.6 Hydrogen plant foundations.

X	y
Qty (cy)	MH
9655	17,678
2416	5582
275	758
109	418
185	712
COVAR (R1, R2)	24,195,791.70
VARP (R2)	13,444,551.49
SLOPE (R1, R2)	1.7997
INTERCEPT (R1, R2)	479.9269

CORREL (R1, R2) = correlation coefficient	0.9983
CORREL (R1, R2)^2 = coefficient determination	0.9967

The coefficient of determination is $R^2 = 0.9967$ and the correlation coefficient, $R = 0.9983$, is a strong indication of correlation. A percentage of 99.8 of the total variation on Y can be explained by the linear relationship between X and Y (described by the regression equation; $Y = 1.799x + 479.93$). The relationship between X and Y variables is such that as X increases, Y also increases.

Appendix A

Statistical and mathematical formulas

Statistical formulas for the mean, variance, and standard deviation

Mean: ybar = y1 + y2 + $\cdots$ + yn; Σ y/n

Variance: S^2 = (y1 − Ybar)^2 + (Y2 + Ybar^2) + $\cdots$ + (Yn − Ybar)^2/n − 1

$$s^2 = \Sigma(yi - ybar)^2/(n-1)$$

Standard   deviation: S = [(y1 − Ybar)^2 + (Y2 + Ybar^2) + $\cdots$ + (Yn − Ybar)^2/n − 1]$^{1/2}$

$$s = [\Sigma(yi - ybar)^2/(n-1)]^{1/2}$$

Straight line graph: Handle and install large bore standard pipe

$$y = a + bx; \quad Y = a + (y - y1)/(x - x1) \ (x)$$

where y is the dependent variable, a is the intercept value along the y axis at x = 0, b is the slope, or the length of the rise divided by the length of the run; b = (y − y1)/(x − x1), x is the independent or control variable.

Mathematical expectation: E(X) = p1X1 + p2X2 + $\cdots$ + pkXk = ΣpX

Normal distribution: Y = $1/(\sigma(2pi) \wedge 1/2)e \wedge - 1/2(X-\mu)^2/\sigma^2$

where μ is the mean, σ is the standard deviation, pi = 3.1416......, e = 2.71828....

Standard form: Y = $1/(2pi \wedge 1/2)e \wedge - 1/2(z^2)$

z is normally distributed with mean 0 and variance 1.

Central limit theorem: W_i = (xbar)$_i$ − $\mu/(\sigma/k^{1/2})$

N (0,1) in the limit as k approaches infinity.

Method of least squares

Least square line

The least square line approximating the set of points (x1, y1), (x2, y2), ..., (xn, yn) has the equation

$$y = bx + a$$

where b is the slope of the line, and a is the y-intercept.

The best fit line for the points (x1, y1), (x2, y2), ..., (xn, yn) is given by

$$y - ybar = b(x - xbar)$$

where the slope is

$$b = \Sigma(xi - xbar)(yi - ybar)/\Sigma(xi - xbar)^2$$

and the y-intercept is

$$a = ybar - bxbar$$

Formula correlation coefficient r: $r = n$ $(\Sigma$ $XY) - (\Sigma X)$ $(\Sigma Y)/[n$ $(\Sigma X^2) - (\Sigma X)^2]$ $[n$ $(\Sigma y^2) - (\Sigma Y)^2]^{1/2}$

Define the U model: $Hn = H1(n^b)$ where Hn is the hours required for the nth unit of production and H1 is the hours required for the first unit.

Natural slope b is defined by the formula: $S = 10^b \log (2) + 2$ logarithm to base 10.

Prediction for the total hours for a "block" of production

Define man hours for a block of erection as the total man hours required to erect all units from unit M to another unit N, $N > M$

TM, N is defined as:

$$TM, N = H1[M^b + (M + 1)^b + (M + 2)^b + \cdots\cdots + N^b]$$

Approximation formula:

$$TM, N = [H1/(1 + b)][(N + 0.5)^{(1 + b)} - (M - 0.5)^{(1 + b)}]$$

Linear regression—fitting U model to unit historical data

$$y = ax^b$$

where y is the hours required for the nth unit of production, a is the hours required for the first unit, and b is the natural slope.

The power function $y = ax^b$ is transformed from a curved line on arithmetic scales to a straight line on log−log scales, let:

$$y = \log y$$
$$a = \log a$$
$$x = \log x$$

taking logarithms of both sides, $\log y = \log a + x \log b$, appears like, $y = a + bx$

Calculating sample size
Sample size: absolute precision

$$n = z^2 \times p \times (1 - p)/e^2$$

where z is the z value (1.96 for 95% confidence level), p is the percentage expressed as a decimal (0.2), and e is the acceptable error percentage as a decimal (0.05 = ± 5%).

Calculating error limits for a sample size
Given sample size is calculated by:

$$n = (z^2 \times p \times (1 - p))/e^2$$

Determine the limit of error, e

$$e = z \times (p \times (1 - p)/n)\hat{1}/2$$

Work sampling method
Model: $Hs = (N_i)(Ht)(RF)(1 + PF\&D)/N$ where Hs is the standard man hours per task; N_i is the observation of event I; Ht is the total man-hours worked during sample study; RF is rating factor; PF&D is the personal, fatigue, and delay allowance; N is the number of random observations during sample study.

Expected-value method
If X denotes a discrete random variable that can assume the values X1, X2, ..., X_i with respective probabilities p1, p2, ..., p_i where $p1 + p2 + \cdots + p_i = 1$, the mathematical expectation of X or simply the expectation of X, denoted by E(X) is defined as $E(X) = p1 \ X1 + p2 \ X2 + \cdots + p_i \ X_i = \Sigma pj \ Xj = \Sigma p \ X$ where E(X) is the expected value of the estimate for event I, pj is the probability that X takes on value Xj, $0 < = Pj (Xj) < = 1$, and Xj is the event.

Range method
The mean and variance for each of the three single cost elements are calculated as

$$E(C_i) = (L + 4M + H)/6$$

$$var(C_i) = ((H - L)/6)\hat{2}$$

where $E(C_i)$ is the expected cost of distribution i, i = 1, 2, n; L is the lowest cost, or best case estimate of cost distribution; M is the modal value, or most likely estimate of cost distribution; H is the highest cost, or worst case estimate of cost distribution; and var (C_i) is the variance of cost distribution i, I = 1,2, ..., n, dollars^2.

The mean of the sum is the sum of the individual means, and the variance is the sum of the variances.

$$E(Cr) = E(C1) + E(C2) + \cdots + E(Cn)$$

$$var(Cr) = var(C1) + var(C2) + \cdots + var(Cn)$$

Where $E(Cr)$ is the expected total cost of independent subdistributions i, and var (Cr) is the variance of total cost of independent subdistributions i.

The probability is calculated using

$$Z = UL - E(Cr)/[var(Cr)]^{1/2}$$

where Z is the value of the standard normal distribution; UL is the upper limit of cost, arbitrarily selected.

Expected profit is defined as:

$$profit = Bp - Ce$$

expected profit $= P \times (Bp - Ce)$ where Bp is the bid price, Ce is the estimated cost, P is the probability of event $(Bp - Ce)$, $0 \leq Prob \leq 1$.

Capture rate
The capture rate is defined as:

$$capture\ rate = (Cs)/(C_i) \times 100,\ percent$$

where Cs is the cost estimates that are successful.

Moving averages.
Smoothing of time series
Given a set of numbers

$$Y1, Y2, \ldots$$

define a moving average of order N to be given by the sequence of arithmetic means,
$$Y1 + Y2 + \cdots + Ny/n, \quad Y2 + Y3 + \cdots + yn/n, \quad Y3 + Y4 + \cdots + yn/n, \ldots$$
The sums in the numerators are moving totals of order n.

Estimation of moving averages
The average of n most recent observations, computed at time t, is given by:

$$Ma = \frac{(yt + yt - 1 + \cdots + yt - n + 1)}{n}$$

where Ma is the moving average of response variable; y is the data, labor, cost, price, etc.; t is the unit of time, years, months, etc.; and n is the denominator of group of time units.

Exponential smoothing
Formula:

$$(D \times S) + (F \times (1 - s))$$

where D is the most recent period's demand, S is the smoothing factor in decimal form, and F is the most recent period's forecast.

Cost index
The current cost is found by using the formula:

$$C = H(Ic/Ih)$$

where C is the current or future cost, H is the historical or past cost, Ic is the index corresponds to current or future time period, and Ih is the index corresponds to historical or past time period.

Appendix B

Excel functions and mathematical functions

Excel functions

Graphic analysis of data

Use Excel's chart capabilities to plot the graphical straight line given by the equation $y = a + bx$ to use the Excel chart capabilities, highlight the range x:y, and select *Insert,* and select from *Charts, Scatter,* go to quick access bar and select from *Chart Tools, Design* and from *Chart Layouts,* Select *Layout 9.*

Excel functions

Excel Statistical Functions for forecasting the value of y for any x. Thus a and b can be calculated in Excel. Where R1 = the array of y values and R2 = the array of x values.

b = SLOPE (R1, R2) = COVAR (R1, R2)/VARP (R2)

a = INTERCEPT (R1, R2) = AVERAGE (R1) − b * Average (R2)

SLOPE (R1, R2) = slope of regression line

INTERCEPT (R1, R2) = y − intercept of the regression line

FORECAST (x, R1, and R2) calculates the predicted value of y for given value of x. Thus

FORECAST (x, R1, and R2) = a + b * x where a = INTERCEPT (R1, R2) and b = SLOPE (R1, R2)

TREND (R1, R2) = array function that produces an array of predicted y values corresponding to x values stored in array R2, based on the regression line calculated from x values stored in array R2, and y values stored in array R1.

COVAR (R1, R2) = returns covariance, the average of the products of deviations for each data point pair in two data cells

VARP (R2) = calculates variance based on the entire population (ignores logical values and text in the population)

Correlation

CORREL (R1, R2) = correlation coefficient of data in arrays R1 and R2

CORREL (R1, R2)^2 = coefficient determination

The mean, variance, and standard deviation measures of central tendency

AVERAGE (number 1, number 2): Returns the average (arithmetic mean) of its arguments, which can be numbers or names, arrays, or references that contain numbers.

VAR (number 1, number 2): Estimates variance based on a sample (ignores logical values and text in the text in the sample

STDEV (number 1, number 2): Estimates standard deviation based on a sample (ignores logical values and text in the sample)

Math & Trig functions

Math formulas

To use the *math formulas*, go to quick access toolbar; select *Math & Trig*; and then select *SUMPRODUCT*.

SUMPRODUCT = Returns the sum of the products of corresponding ranges or arrays. Arrays 1, 2, and 3 are 2−255 arrays for which you want to multiply and then add components. All arrays must have the same dimensions.

LOG = Returns the logarithm of a number to the base you want the logarithm.

Number is the positive real number for which you want the logarithm.

SQRT = Returns the square root of a number. Number is the number for which you want the square root.

Appendix C

Area and volume formulas

Formulas—areas and volumes
Square: area = $(edge)^2$; $A = a^2$
Rectangle: base $\times$ altitude; $A = ba$
Right triangle: Area = 1/2 base $\times$ altitude; $A = 1/2\ ba$
Pythagorean theorem:
$(Hypotenuse)^2$ + sum of squares of two legs of right triangle
$C^2 = a^2 + b^2$; $a = [c^2 - b^2 1/2]$
Oblique triangle; Area = 1/2 base $\times$ altitude; $A = 1/2\ bh$
$A = [s(s - a)(s - b)(s - c)]\wedge 1/2$, where $s = (a + b + c)/2$
Parallelogram; opposite sides are parallel; $A = bh$
Trapezoid; one pair of opposite sides parallel
$A = 1/2$ sum of bases $\times$ altitude; $A = 1/2\ (a + b)\ h$
Circle: circumference = 2 (pi)(radius) = (pi) (diameter; $C = 2(pi)R = (pi)D$
Area = (pi) $(radius)^2$ = (pi/4) $(diameter)^2$: $A = (pi)R^2 = (pi/4)D^2$
Sector of circle; area = 1/2 radius $\times$ arc; $A = 1/2\ Rc = 1/2\ R^2$ angle
Segment of circle; area (segment) = area (sector) $-$ area(triangle)
$A = Rc - 1/2\ ba$
Ellipse; area = (pi)ab
Parabolic segment; area = 2/3 ld
Right circular cone; $V = (pi)r^2 h$; A = side area + base area;
$A = (pi)r[r + (r^2 + h^2)\wedge 1/2]$
Right circular cylinder; $V = (pi)r^2 h = (pi)d^2 h/4$
A = side area + end areas = 2 (pi)r (h + r)

Appendix D

Standard to metric

Lengths

Metric conversion
1 centimeter = 10 millimeters; 1 cm = 10 mm
1 meter = 100 centimeters; 1 m = 100 cm
Standard conversions
1 foot = 12 inches; 1 ft. = 12 in.
1 yard = 3 feet; 1 yd = 3 ft.
1 yard = 36 inches; 1 yd = 36 in.
Metric−standard conversions
1 millimeter = 0.03937 inches; 1 mm = 0.03937 in.
1 centimeter = 0.39370 inches; 1 cm = 0.39370 in.
1 meter = 39.39008 inches; 1 m = 39.37008 in.
1 meter = 3.28084 feet; 1 m = 3.28084 ft.
1 meter = 1.093.6133 yards; 1 m = 1.0993.6133 yd
Standard−metric conversions
1 inch = 2.54 centimeters; 1 in. = 2.54 cm
1 foot = 30.48 centimeters; 1 ft. = 30.48 cm
1 yard = 91.44 centimeters; 1 yd = 91.44 cm
1 yard = 0.9144 meters; 1 yd = 0.3144 m

Volumes

Metric conversion
1 cubic centimeter = 1000 cubic millimeters; 1 cu cm = 1000 cu mm
1 cubic meter = 1 million cubic centimeters; 1 cu m = 1,000,000 cu cm
Standard conversions
1 cubic foot = 1728 cubic inches; 1 cu ft. = 1728 cu in.
1 cubic yard = 46,656 cubic inches; 1 cu yd = 46,656 cu in.
1 cubic yard = 27 cubic feet; 1 cu yd = 27 cu ft.
Metric−standard conversions

1 cubic centimeter = 0.06102 cubic inches; 1 cu cm = 0.06102 cu in.
1 cubic meter = 35.31467 cubic feet; 1 cu m = 35.31467 cu ft.
1 cubic meter = 1.30795 cubic yards; 1 cu m = 1.30795 cu yd
Standard−metric conversions
1 cubic inch = 16.38706 cubic centimeters; 1 cu in. = 16.38706 cu cm
1 cubic foot = 0.02832 cubic meters; 1 cu ft. + 0.02832 cu m
1 cubic yard = 0.76455 cubic meters; 1 cu yd = 0.76455 cu m

Areas

Metric conversion
1 sq centimeter = 100 sq millimeters; 1 sq cm = 100 sq mm
1 sq meter = 10,000 sq centimeters; 1 sq m = 10,000 sq cm
Standard conversions
1 sq foot = 144 sq inches; 1 sq ft. = 144 sq in.
1 sq yard = 9 sq feet; 1 sq yd = 9 sq ft.
Metric−standard conversions
1 sq centimeter = 0.15500 sq inches; 1 sq cm = 0.15500 sq in.
1 sg meter = 10.76391 sq feet; 1 sq m = 10.76391 sq ft.
1 sq meter = 1.19599 sq yards; 1 sq m = 1.9599 sq yd
Standard−metric conversions
1 sq inch = 6.4516 sq centimeters; 1 sq in. = 6.4516 sq cm
1 sq foot = 929.0304 sq centimeters; 1 sq ft. = 929.0304 cm
1 sq foot = 0.09290 sq meters; 1 sq ft. = 0.09290 sq m

Appendix E

Boiler man hour tables

Wachs Trav-L-Cutter—use
0.033−0.050 MH/in. of circumference for every 0.5″ of wall thickness to be removed (Table E.1).

Mill with power tool—use
0.05−0.10 MH/in. of circumference for every 0.5″ of wall thickness to be removed.

Saw or grind—use mill with power tool rate
Formula: Circumference in inches × rate/inch) × no of passes (0.5″ cut per pass)

Example: 22″ diameter × 2″ wall
[(22′ diameter × 3.1416) × 0.033/in.] (2″ wall 1/2) = 9.1 hours

$$\underset{\text{Circumference}}{\qquad} \quad \underset{\text{Rate}}{\qquad} \quad \underset{\text{No. of passes}}{\qquad}$$

Rates for carbon steel only, double for all other material
Excludes overhead labor, set-up, maintenance, and removal of equipment (Tables E.2 and E.3).

TABLE E.1 Cutting and milling rates for 0−5000 PSI.

Size (OD)	Torch Cut	Saw or Grind	Mill with Power tool
0″ < diameter ≤ 3″	0.15−0.20	0.35−0.50	0.35−0.50
3″ < diameter ≤ 4.5″	0.18−0.30	0.375−1.00	0.375−1.00
4.5″ < diameter ≤ 6.5″	0.25−0.60	0.50−3.00	0.50−3.00
6.5″ < diameter ≤ 35″ with WT from 0.5″ thru 5.5″	Torch cut		

TABLE E.2 Expanding rates, MH per tube end.

PSI	2″	2.5″	3″	4″
160	0.19	0.22	0.44	0.94
200	0.20	0.24	0.45	0.95
300	0.22	0.27	0.47	0.96
400	0.25	0.31	0.49	0.97
500	0.28	0.36	0.50	0.99
600	0.30	0.40	0.53	1.02
700	0.36	0.43	0.59	1.07
800	0.36	0.43	0.66	1.33
900	0.36	0.43	0.72	1.60
1000	0.36	0.43		1.87
1100	0.50			
1200	0.50			
1300	0.50			
1400	0.50			
1500	0.50			
1600	0.50			
1700	0.50			
1800	1.20			
1900	1.20			
2000	1.20			
2100	1.20			

For seal welding 2″ generating tubes inside drums, reexpanding, and NDE, use 1.2 MH/JT (Table E.4).

Field welding of tubes in the heat input zones of all boilers of 2000 psi design and over is by the TIG process.

Field welding of tubes by the SMAW process; root pass by the TIG process PWHT:

1. Carbon steel > 0.75″ thick
2. Chrome molly steel with carbon content >0.25% and wall thickness >0.50″
3. Croloy materials with more than 3% chromium, or diameter >4″ and WT > 0.50″ (Table E.5).

TABLE E.3 Socket and seal welding, MH per weld.

MH per weld	Seal welding			Socket welding	
	Outside	Inside		W/O stress	W/stress
Size	Header	Header		Relieving	Relieving
1" O.D. tube	0.7	0.9		1.1	1.4
Over 1"–1.5" O.D. tube	0.9	1.1		1.2	1.5
Over 1.5"–2" O.D. tube	1.2	1.3		1.4	1.7
Over 2"–2.5" O.D. tube	1.4	1.5		1.7	2.2
Over 2.5"–3.25" O.D. tube	1.7	1.8		2.1	2.6
Over 3.25"–4" O.D. tube	1.9	2.1			
Over 4"–4.5" O.D. tube	2.1	2.4			
Over 4.5"–5.5" O.D. tube	2.3	2.7			
Place and weld					
2.5" H.H. fitting	2.2				
3.25" H.H. fitting	2				
4" H.H. fitting or blind nipple	2.5				
4.5" H.H. fitting or blind nipple	2.7				
Radiograph and header end plugs	2.8				

TABLE E.4 Field tube welding, MH/weld.

| | Tube welding—MH Per weld | | | | | |
| | Design pressure (PSI) | | | | | |
	Up To 500	501 To 1000	1001 To 1500	1501 To 2000	2001 To 2500	2501 To 3000
Tube size (OD)						
1"<BW ≤1.5" TIG	2.4	2.6	2.8	3.0	3.2	3.4
1.5"<BW ≤2" TIG	2.7	2.9	3.4	3.7	4.0	4.3
2"<BW ≤2.5" TIG	3.2	3.3	3.8	4.2	4.2	4.4
2.5"<BW ≤3" TIG	3.7	4.0	4.1	4.8	4.6	4.8
3"<BW ≤3.5" TIG	4.3	4.6	4.8	5.3	5.3	5.6
3.5"<BW ≤4" TIG	4.9	5.2	5.4	5.9	5.9	6.2
4"<BW ≤4.5" TIG	5.5	5.8	6.1	6.5	6.5	6.7
4.5"<BW ≤5.5" TIG	6.8	7.1	7.4	7.7	8.0	8.3
4.5"<ring weld ≤5.5" SMAW PWHT	8.4	9.4	10.3	11.5	12.4	13.6
5.5"<ring weld ≤6.5" SMAW	5.0	5.8	6.4	6.8	7.0	7.2
5.5"<BW ≤6.5" TIG	7.7	8.0	8.5	9.0	9.5	10.0
5.5"<ring weld ≤6.5" SMAW PWHT	10.2	11.0	12.0	13.8	14.2	15.0

TABLE E.5 Field pipe welding, MH per weld.

| | | | | | | Wall thickness in inches | | | | | | | |
Diameter inches	0.250	0.500	0.750	1.000	1.250	1.500	1.750	2.000	2.250	2.500	2.750	3.000
6	3.0	3.3	7.2	8.7	13.2	15.0	16.2	17.4				
8	4.0	4.4	9.6	11.6	17.6	20.0	21.6	23.2	24.8	28.0		
10	5.0	5.5	12.0	14.5	22.0	25.0	27.0	29.0	31.0	35.0	38.0	43.0
12	6.0	6.6	14.4	17.4	26.4	30.0	32.4	34.8	37.2	42.0	45.6	51.6
14	7.0	7.7	16.8	20.3	30.8	35.0	37.8	40.6	43.4	49.0	53.2	60.2
16	8.0	8.8	19.2	23.2	35.2	40.0	43.2	46.4	49.6	56.0	60.8	68.8
18	9.0	9.9	21.6	26.1	39.6	45.0	48.6	52.2	55.8	63.0	68.4	77.4
20	10.0	11.0	24.0	29.0	44.0	50.0	54.0	58.0	62.0	70.0	76.0	86.0
22	11.0	12.1	26.4	31.9	48.4	55.0	59.4	63.8	68.2	77.0	83.6	94.6
24	12.0	13.2	28.8	34.8	52.8	60.0	64.8	69.6	74.4	84.0	91.2	103.2
26	13.0	14.3	31.2	37.7	57.2	65.0	70.2	75.4	80.6	91.0	98.8	111.8
28	14.0	15.4	33.6	40.6	61.6	70.0	75.6	81.2	86.8	98.0	106.4	120.4
30	15.0	16.5	36.0	43.5	66.0	75.0	81.0	87.0	93.0	105.0	114.0	129.0
32	16.0	17.6	38.4	46.4	70.4	80.0	86.4	92.8	99.2	112.0	121.6	137.6
34	17.0	18.7	40.8	49.3	74.8	85.0	91.8	98.6	105.4	119.0	129.2	146.2
35	17.5	19.3	42.0	50.8	77.0	87.5	94.5	101.5	108.5	122.5	133.0	150.5

PWHT:

1. Carbon steel $> 0.75''$ thick
2. Chrom moly steel with carbon content $>0.25\%$ and wall thickness $>0.50''$
3. Croloy materials with more than 3% chromium, or diameter $>4''$ and WT $> 0.50''$ (Table E.6).

TABLE E.6 Diamond soot blowers, MH per SB W/PVF.

Type of unit	Manual Operation	MH	Unit
G-2, G-21, G-9B, A2E	SB	22	EA
	SB W/PVF	44	EA
	Total	66	EA
IR and 1S	SB	20	EA
	SB W/PVF	44	EA
	Total	64	EA
1K, T9, and T11	SB	35	EA
	SB W/PVF	51	EA
	Total	86	EA
G9B, A2E (swinging arm)	SB	18	EA
	SB W/PVF	51	EA
	Total	69	EA
1K, DE2, SE2 (strait line)	SB	100	EA
	SB W/PVF	51	EA
	Total	151	EA
Pressure reducing station		70	EA
Trays and channels for supporting tubing		18	EA
Tubing for automatic sequential air operation		40	EA
Air compressor and/or receiver		18	ton
Auto sequential panel or air master controller		86	Panel
IK structural supports		6	Blower

TABLE E.7 Structural steel.

Structural steel and miscellaneous iron

Erect structural steel

Main steel	MH	Unit
Erect structural steel; ≤ 20 ton		
Light—0–19 lb/ft.	28.0	ton
Medium—20–39 lb/ft.	24.0	ton
Heavy—40–79 lb/ft.	20.0	ton
X heavy—80–120 lb/ft.	16.0	ton
Erect structural steel; 20 ton > tons ≤ 100 ton		
Light—0–19 lb/ft.	21.0	ton
Medium—20–39 lb/ft.	18.0	ton
Heavy—40–79 lb/ft.	15.0	ton
X heavy—80–120 lb/ft.	12.0	ton
Erect structural steel; >100 ton		
Light—0–19 lb/ft.	16.8	ton
Medium—20–39 lb/ft.	14.4	ton
Heavy—40–79 lb/ft.	13.2	ton
X heavy—80–120 lb/ft.	11.8	ton
PLATFORM FRAMING	0.15	SF
HANDRAIL & TOE PLATE	0.25	LF
FLOOR GRATING	0.20	SF
STAIR TREADS	0.85	LF
STRAIGHT LADDER	0.30	LF
CAGED LADDERS	0.35	LF
Ladders and safety cage—KD	1.00	LF
Girths (side wall support steel for metal siding)	28.00	ton
Purlins (roof support steel for metal siding)	28.00	ton
Elevator steel	38.00	ton

For combustion steam and air blowing soot blowers, double PVF man hours.

For Vulcan soot blowers, add 24 MH/soot blower for scavenger drain PVF

TABLE E.8 Burners.

Description	MH	Unit
Circular burners	32.00	ton
Air jet low NO_x coal burner	180.00	EA
DRB-4Z low NO_x coal burner	240.00	EA
XLC low NO_x oil and gas burner	240.00	EA
Dual zone NO_x port (over fire air system)	200.00	EA
Inter tube burners, includes blocks, tips, and riffle casings	18.00	ton
Shop assembled burners in wind box, includes welding	24.00	ton
Automatic lighters—includes welding	6.00	EA
Oil atomizers	4.00	EA

Removal of soot blower use 50% of table man hours (Tables E.7 and E.8).

Index

Printed in the United States
By Bookmasters